CMOS Gate-Stack Scaling —
Materials, Interfaces and
Reliability Implications

MATERIALS RESEARCH SOCIETY
SYMPOSIUM PROCEEDINGS VOLUME 1155

CMOS Gate-Stack Scaling — Materials, Interfaces and Reliability Implications

Symposium held April 14–16, 2009, San Francisco, California, U.S.A.

EDITORS:

Alexander A. Demkov
The University of Texas at Austin
Austin, Texas, U.S.A.

Bill Taylor
Sematech, Inc.
Albany, New York, U.S.A.

H. Rusty Harris
Texas A&M University
College Station, Texas, U.S.A.

Jeffery W. Butterbaugh
FSI International
Chaska, Minnesota, U.S.A.

Willy Rachmady
Intel Corporation
Hillsboro, Oregon, U.S.A.

Materials Research Society
Warrendale, Pennsylvania

CAMBRIDGE
UNIVERSITY PRESS

32 Avenue of the Americas, New York NY 10013-2473, USA

Cambridge University Press is part of the University of Cambridge.

It furthers the University's mission by disseminating knowledge in the pursuit of education, learning and research at the highest international levels of excellence.

www.cambridge.org
Information on this title: www.cambridge.org/9781605111285

Materials Research Society
506 Keystone Drive, Warrendale, PA 15086
http://www.mrs.org

© Materials Research Society 2009

First published 2009
First paperback edition 2012

Single article reprints from this publication are available through University Microfilms Inc., 300 North Zeeb Road, Ann Arbor, MI 48106

CODEN: MRSPDH

A catalogue record for this publication is available from the British Library

ISBN 978-1-605-11128-5 Hardback

*Invited Paper

*Invited Paper

ALTERNATE CHANNEL MATERIALS

PREFACE

This proceedings volume contains a selection of papers presented at Symposium C, "CMOS Gate-Stack Scaling — Materials, Interfaces and Reliability Implications," held April 14–16 at the 2009 MRS Spring Meeting in San Francisco, California. To address the demands of device scaling, new materials are being introduced into conventional Si CMOS processing at an unprecedented rate. Articles collected in this book focus on understanding, from a chemistry and materials perspective, the mechanism of interface formation and defects at interfaces, for both conventional Si and alternative channel (Ge or III-V) systems. Several papers address reliability concerns for HiK/metal gate (basic physical models, charge trapping, etc.), while others cover characterization of the thin films and interfaces which comprise the gate stack.

As symposium organizers, we are indebted to all the participants who made the symposium an exciting event, and in particular, to all the authors who contributed papers to this volume. We would like to acknowledge the generous financial support of DCA Instruments, FSI International and one anonymous donor.

Alexander A. Demkov
Bill Taylor
H. Rusty Harris
Jeffery W. Butterbaugh
Willy Rachmady

September 2009

MATERIALS RESEARCH SOCIETY SYMPOSIUM PROCEEDINGS

MATERIALS RESEARCH SOCIETY SYMPOSIUM PROCEEDINGS

Prior Materials Research Society Symposium Proceedings available by contacting Materials Research Society

Advanced Si-based Gate Stacks

Mater. Res. Soc. Symp. Proc. Vol. 1155 © 2009 Materials Research Society   1155-C12-01

Defects in HfO$_2$ Based Dielectric Gate Stacks

Patrick M. Lenahan[1], Jason T. Ryan[1], Corey J. Cochrane[1], and John F. Conley Jr.[2]

[1]The Pennsylvania State University, University Park, PA 16802, U.S.A.

[2]Oregon State University, Corvallis, OR 97331, U.S.A.

ABSTRACT

We report on both conventional electron paramagnetic resonance (EPR) measurements of fully processed HfO$_2$ based dielectric films on silicon and on electrically detected magnetic resonance (EDMR) measurements of fully processed HfO$_2$ based MOSFETs. The magnetic resonance measurements indicate the presence of oxygen vacancy and oxygen interstitial defects within the HfO$_2$ and oxygen deficient silicons in the interfacial layer. The EDMR results also indicate the generation of at least two defects when HfO$_2$ based transistors are subjected to significant negative bias at modest temperature. Our results indicate generation of multiple interface/near interface defects, likely involving coupling with nearby hafnium atoms.

INTRODUCTION

Tremendous progress has been made towards replacing conventional (SiO$_2$ and SiO$_x$N$_y$) gate dielectrics with HfO$_2$ based materials in high performance metal-oxide-silicon (MOS) field effect transistors (MOSFETs) [1, 2]. However, there are many challenges in the integration of these materials into MOS technology. Critical issues include: trapping in HfO$_2$ [3, 4], trapping in the interfacial layer dielectric between the silicon and the HfO$_2$ [5, 6], silicon/dielectric interface traps [1] and various instabilities which may involve either the generation of new traps or the population of existing traps or both of these processes [7].

Magnetic resonance techniques, such as conventional electron paramagnetic resonance (EPR) or electrically detected magnetic resonance (EDMR) techniques such as spin dependent recombination (SDR), are the most powerful tools currently available to identify the structure of these trapping centers [8]. The technique is generally sensitive to paramagnetic defects. (Nearly all electrically active defects can be rendered paramagnetic.) Information about the local structure of trapping defects can be gleaned from the relationship between the magnetic field and microwave frequency at which resonance occurs. In the simplest of cases, the resonance condition is given by [8],

$$h\nu = g\beta H, \qquad (1)$$

where h is Planck's constant, ν is the microwave frequency, β is the Bohr magneton, H is the magnetic field and g is number, typically close to 2, which depends upon the relationship between the defect's orientation and the magnetic field vector. The g is expressed in terms of a matrix which is often called the g tensor. (This simple relationship is often altered by the presence of nearby magnetic nuclei and other factors which will not be particularly relevant for the results presented in this paper.)

HfO₂ DEFECTS

At least two intrinsic defects have been observed in magnetic resonance observations of HfO$_2$ films on silicon: oxygen vacancies and oxygen interstitials. Kang et al.[9] and Ryan et al [10]. reported the generation of two strong signals in HfO$_2$ films subjected to, respectively, vacuum ultraviolet (VUV) illumination under bias and gamma irradiation. (Ryan et al. [10] also reported the generation of P$_b$ like Si/dielectric interface traps.) Figure 1 illustrates a post- VUV EPR trace and figure 2 illustrates pre- and post- gamma irradiation traces. (The spectrometer settings are slightly different in the VUV and gamma irradiation traces; in addition, the generation of P$_b$ centers in the gamma irradiation case somewhat alters the results of figure 2.) These representative results indicate the presence of oxygen vacancies and oxygen interstitials [9-11]. Figure 3 illustrates a comparison of a post VUV trace and a simulated EPR trace for a defect with g matrix components corresponding to $g_{zz} = 2.04$, $g_{yy} = 2.01$, and $g_{xx} = 2.000$. These g matrix components correspond closely to the well established components for oxygen interstitials observed previously in quite similar materials such as ZrO$_2$. The signal on the far right of both traces were tentatively linked to an oxygen vacancy by Ryan et al. [10] and by Lenahan and Conley [11]. This somewhat tentative identification became more definitive with recent density functional theory calculations by Ramo et al. [12] who predicted that a (positively charged) oxygen vacancy would have a g matrix corresponding to $g_{xx} = 1.9835$, $g_{yy} = 1.963$, and $g_{zz} = 1.9450$. Figure 4 illustrates a comparison of the right side signal trace in figures 1 and 2 with that of a calculated trace utilizing the Bruker Biospin SimFonia software and the g matrix components calculated by Ramo et al. [12] The higher broadening trace is an excellent match to the experimental data. (The low broadening trace indicates that, at the very least, the center of the spectrum and its approximate width agree with the Ramo et al. calculations.) Figure 5 illustrates a comparison of the data from figures 1 and 2 and a simulation of the sum of the oxygen interstitial and oxygen vacancy centers, again utilizing the Bruker Biospin SimFonia software. The agreement between the simulation and the data is quite close. The observations of oxygen vacancies and oxygen interstitials as dominating HfO$_2$ defects is generally consistent with several "theoretical" studies of HfO$_2$ defects [13, 14].

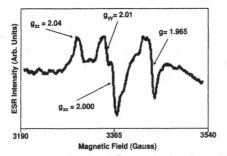

Figure 1. An EPR trace taken on an HfO$_2$ film on silicon which has been subjected to vacuum ultraviolet illumination hc/λ <10.2eV under negative gate bias. The illumination resulted in an electron fluence of approximately 2×10^{13} cm^{-2} in the "bulk" of the film and a comparable hole fluence near its surface. The trace indicates the presence of high densities (> 10^{12} cm^{-2}) of oxygen vacancies and oxygen interstitials.

4

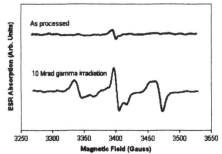

Figure 2. Pre-irradiation (above) and post-irradiation (below) wide scan ESR traces indicating the generation of several defects in the $Hf(NO_3)_4$ precursor HfO_2 dielectric film on H-terminated silicon. The two peaks on the left are (mostly) due to an O_2^- coupled to a hafnium ion (the central peak includes a small contribution from P_b centers). The peak on the far right is likely due to an oxygen vacancy in the HfO_2. In these traces, the spectrometer settings have been set to optimize the O_2^- and oxygen vacancy spectra. The post irradiation trace indicates the presence of high densities ($>10^{12}$ cm^{-2}) of both oxygen vacancies and oxygen interstitials.

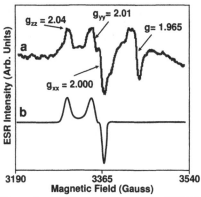

Figure 3. (a) EPR trace generated by VUV illumination under bias compared to the (b) simulated EPR spectra with $g_{xx} = 2.04$, $g_{yy} = 2.01$, and $g_{zz} = 2.000$.

5

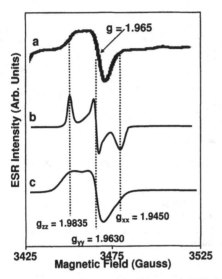

Figure 4. (a) EPR data from Ryan et al. [10], (b) positive oxygen vacancy spectrum simulation with low broadening, and (c) positive oxygen vacancy spectrum simulation with high broadening. In both simulations, the g matrix calculations of Ramo et al. [12] were utilized.

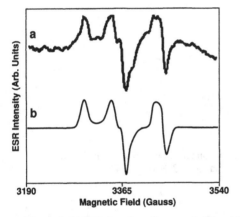

Figure 5. (a) EPR data from Kang et al. [9] and (b) simulation of the positive oxygen vacancy and oxygen interstitial. The simulation is based on the sum of simulations in figures 3 and 4.

DEFECTS IN THE INTERFACIAL LAYER

Ryan et al. [15, 16] and Bersuker et al. [17] have reported the observation of several E' center variants in the interfacial layer of HfO_2 based structures. (E' centers are oxygen deficient silicons, dominating deep level defects in conventional SiO_2 based oxides. Their EPR spectra are characterized by narrow lines with zero crossing g values between 2.000 and 2.003.) Ryan et al. [15, 16] and Bersuker et al. [17] noted a very strong processing dependence of these defects in the interfacial layer. Figures 6 and 7 illustrate this strong process dependence.

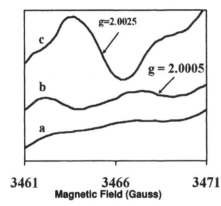

3461   **3466**   **3471**
Magnetic Field (Gauss)

Figure 6. Comparison of three very differently processed samples; (a) 3 nm HfO_2 deposited on etch-back in-situ steam generated SiO_2, (b) a sample identical to (a) except that it received a 700 and 1000°C N_2 post deposition anneal, and (c) a sample identical to (b) except that it had a 10nm thick TiN cap deposited prior to N_2 annealing.

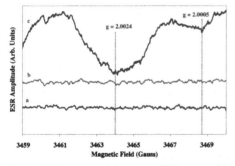

Figure 7. Three second derivative EPR traces of (a) control 1nm ozone grown SiO_2, (b) 3 nm HfO_2 deposited on 1nm ozone grown SiO_2, and (c) oxygen deficient HfO_2 deposited on 1nm ozone grown SiO_2. Note that the deposition of the oxygen deficient HfO_2 film grossly increases the densities of the two E' like signals.

Ryan et al. [15] also noted that these interfacial E' variants exist in an environment which cannot be fully amorphous or a non textured polycrystalline matrix. They showed that this must be so because the spectrum depends on the orientation of the device structures in the magnetic field. This is illustrated in figure 8.

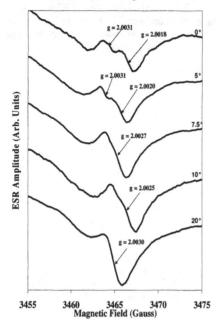

Figure 8. EPR vs. sample orientation. Note the change in line shape as the sample is rotated in the applied magnetic field. This demonstrates that the defect cannot reside in a purely amorphous or non-textured polycrystalline matrix.

Ryan et al. [15] also showed that the amplitude of these E' signals depends upon the position of the silicon/dielectric Fermi energy. This result demonstrates that the centers can act electronically as interface traps. This result is illustrated in figure 9. As the Fermi level passes up and through the silicon/dielectric interface band gap, the centers can pick up first one electron, rendering the center paramagnetic (EPR active) and then a second electron rendering the defect diamagnetic (EPR inactive). Since these centers have also been observed in HfO_2 and hafnium silicate structures by at least two groups [18, 19], sometimes at high density, it is likely that they are performance limiting defects of widespread importance.

8

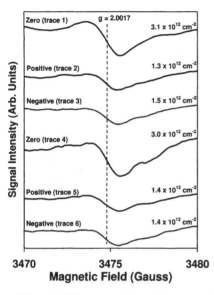

Zero (trace 1)   g = 2.0017 3.1×10^{12} cm^{-2}

Positive (trace 2) 1.3×10^{12} cm^{-2}

Negative (trace 3) 1.5×10^{12} cm^{-2}

Zero (trace 4) 3.0×10^{12} cm^{-2}

Positive (trace 5) 1.4×10^{12} cm^{-2}

Negative (trace 6) 1.4×10^{12} cm^{-2}

Signal Intensity (Arb. Units)

3470   3475   3480
Magnetic Field (Gauss)

Figure 9. ESR vs. applied gate bias. Note the amplitude modulation as a result of biasing. The defect acts as both an electron and hole trap.

INTERFACE/NEAR INTERFACE DEFECTS GENERATED IN BIAS TEMPERATURE INSTABILITIES

Cochrane et al. [20] have reported on the generation of interface/near interface defects in fully processed HfO_2 based transistors subjected to negative bias stressing at room temperature and elevated temperature. The key results are illustrated in figures 10, 11, and 12. Figure 10 illustrates gated diode (DCIV) recombination current measurements on an unstressed transistor, a transistor subjected to brief room temperature negative bias stress, and a more lengthy negative bias stress. The amplitude of the DCIV peak scales with the density of interface traps. Figure 11 illustrates pre- and post- room temperature stressing EDMR measurements obtained by SDR. Figure 12 illustrates pre- and post- high temperature stressing SDR measurements. Note that rather broad SDR spectrum with a different zero crossing g values. The short room temperature stressing SDR response g = 1.9998; the long high temperature stressing SDR response g = 2.0026. The post high temperature stressing SDR spectrum is significantly narrower than the room temperature stressing signal. Figure 13 compares long time high temperature negative bias stressed HfO_2 device response to the response of a similarly stressed plasma nitrided SiO_2 (PNO) device and a SiO_2 device. As reported elsewhere, in SiO_2 devices, the process is dominated by P_b centers [21]; in PNO devices, the process is dominated by K centers [22], silicons back bonded to three nitrogen atoms. The SDR results on the HfO_2 results demonstrate that the NBTI defects in these devices are clearly different than those in conventional SiO_2 and PNO devices.

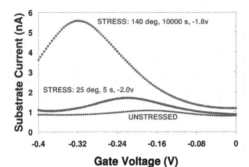

Figure 10. Comparison of DCIV results for an unstressed device, a device subjected to a 5s room temperature stress (25 °C at −2.0 V), and a device subjected to extended stressing at an elevated temperature (140 °C at −1.8 V for 10 000 s).

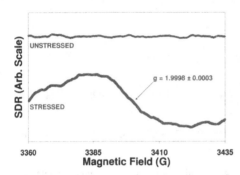

Figure 11. SDR results for an unstressed and 5s room temperature negative bias stressed device (25 °C and −2.0 V). SDR was performed on this device with a gate voltage of −0.220 V. The g of 1.9998 corresponds to a microwave frequency $v = 9.5165$ GHz and magnetic field H=3400 G.

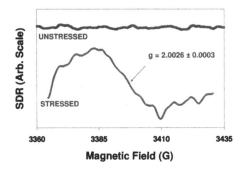

Magnetic Field (G)

Figure 12. SDR results for an unstressed and high temperature stressed (140 °C at −1.8 V for 10,000 s) device. Note that the long term high temperature stressing creates a significantly different SDR response than the brief room temperature stressing. SDR was performed on this device with a gate voltage of −0.310 V. The g of 2.0026 corresponds to a microwave frequency ν =9.5149 GHz and magnetic field H=3394 G.

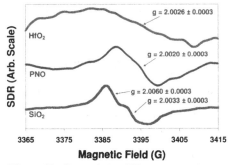

Magnetic Field (G)

Figure 13. Comparison of SDR traces on long term stressed HfO_2, SiO_2, and PNO devices. The g value of the HfO_2 long term stress is very similar to that of the PNO stressed device though it is clear that the signals are not the same. The equivalent oxide thicknesses of the SiO_2 device were 75 and 23 nm for the plasma nitrided device.

The much greater breadth of the HfO_2 signal, especially the SDR response in the short stressing case, suggests some interaction of a nearby hafnium atom with an oxygen deficient silicon. Ryan et al. [15] have, as mentioned previously, also suggested a coupling between an oxygen deficient silicon and a hafnium atom in the interfacial layer. Recently, van Benthem et al. [23] have argued for the presence of hafnium in the interfacial layer.

CONCLUSIONS

Magnetic resonance results indicate considerable complexity is possible in the trapping centers in HfO_2 based gate stacks. We observe two defects in HfO_2 films: an oxygen vacancy and an oxygen interstitial defect. We observe oxygen deficient silicons (E' centers) in the interfacial layer. Several defect center spectra are generated in HfO_2 based devices subjected to negative bias stressing. Although a definitive identification of these negative bias stressing centers is not yet available, they are clearly different from the negative bias stressing centers generated in conventional SiO_2 and plasma nitrided SiO_2 based devices. The considerable breadth of these negative bias stressing induced centers suggests that the defects are coupled to some extent with nearby hafnium atoms, though their line widths and spin lattice relaxation times suggest that they do not involve unpaired electrons directly localized on hafnium atoms. The results collectively demonstrate that magnetic resonance will be a powerful tool in developing a comprehensive understanding of HfO_2 based MOS materials physics.

REFERENCES

1. G.D. Wilk, R.M. Wallace, and J.M. Anthony, *J. Appl. Phys.*, **89**, 5243-5275 (2001).
2. J. Robertson, *Rep. Prog. Phys.*, **69**, 327-396 (2006).
3. G. Ribes, J. Mitard, M. Denais, S. Bruyere, F. Monsieur, C. Parthasarathy, E. Vincent, and G. Ghibaudo, *IEEE Trans. Dev. Mater. Reliab.*, **5**, 5-19 (2005).
4. S. Zafar, A. Kumar, E. Gusev, and E. Cartier, *IEEE Trans. Dev. Mater. Reliab.*, **5**, 45-64 (2005).
5. D. Heh, C.D. Young, G.A. Brown, P.Y. Hung, A. Diebold, E.M. Vogel, J.B. Bernstein, and G. Bersuker, *IEEE Trans. Electron Devices*, **54**, 1338-1345 (2007).
6. C.D. Young, D. Heh, S.V. Nadkarni, C. Rino, J.J. Peterson, J. Barnett, L. Byoung Hun, and G. Bersuker, *IEEE Trans. Dev. Mater. Reliab.*, **6**, 123-131 (2006).
7. K. Onishi, C. Rino, K. Chang Seok, C. Hag-Ju, K. Young Hee, R.E. Nieh, H. Jeong, S.A. Krishnan, M.S. Akbar, and J.C. Lee, *IEEE Trans. Electron Devices*, **50**, 1517-1524 (2003).
8. P.M. Lenahan and J.F. Conley, *J. Vac. Sci. Technol., B*, **16**, 2134-2153 (1998).
9. A.Y. Kang, P.M. Lenahan, and J.F. Conley, *Appl. Phys. Lett.*, **83**, 3407-3409 (2003).
10. J.T. Ryan, P.M. Lenahan, A.Y. Kang, J.F.C. Jr., G. Bersuker, and P. Lysaght, *IEEE Trans. Nucl. Sci.*, **52**, 2272-2275 (2005).
11. P.M. Lenahan and J.F. Conley, *IEEE Trans. Dev. Mater. Reliab.*, **5**, 90-102 (2005).
12. D.M. Ramo, J.L. Gavartin, A.L. Shluger, and G. Bersuker, *Phys. Rev. B*, **75**, 205336 (2007).
13. A.S. Foster, V.B. Sulimov, F.L. Gejo, A.L. Shluger, and R.M. Nieminen, *Phys. Rev. B*, **64**, 224108 (2001).
14. K. Xiong, J. Robertson, and S.J. Clark, *Phys. Status Solidi B-Basic Solid State Phys.*, **243**, 2071-2080 (2006).
15. J.T. Ryan, P.M. Lenahan, J. Robertson, and G. Bersuker, *Appl. Phys. Lett.*, **92**, 123506 (2008).
16. J.T. Ryan, P.M. Lenahan, G. Bersuker, and P. Lysaght, *Appl. Phys. Lett.*, **90**, 173513 (2007).

17. G. Bersuker, C.S. Park, J. Barnett, P.S. Lysaght, P.D. Kirsch, C.D. Young, R. Choi, B.H. Lee, B. Foran, K. van Benthem, S.J. Pennycook, P.M. Lenahan, and J.T. Ryan, *J. Appl. Phys.*, **100**, 094108 (2006).
18. B.B. Triplett, P.T. Chen, Y. Nishi, P.H. Kasai, J.J. Chambers, and L. Colombo, *J. Appl. Phys.*, **101**, 013703 (2007).
19. A. Stesmans and V.V. Afanas'ev, *J. Appl. Phys.*, **97**, 033510 (2005).
20. C.J. Cochrane, P.M. Lenahan, J.P. Campbell, G. Bersuker, and A. Neugroschel, *Appl. Phys. Lett.*, **90**, 123502 (2007).
21. J.P. Campbell, P.M. Lenahan, A.T. Krishnan, and S. Krishnan, *Appl. Phys. Lett.*, **87**, 204106 (2005).
22. J.P. Campbell, P.M. Lenahan, A.T. Krishnan, and S. Krishnan, *J. Appl. Phys.*, **103**, 044505 (2008).
23. K. van Benthem, A.R. Lupini, M. Kim, H.S. Baik, S. Doh, J.H. Lee, M.P. Oxley, S.D. Findlay, L.J. Allen, J.T. Luck, and S.J. Pennycook, *Appl. Phys. Lett.*, **87**, 034104 (2005).

Mater. Res. Soc. Symp. Proc. Vol. 1155 © 2009 Materials Research Society   1155-C09-12

Investigating the Interfacial Properties of Hf-Based/Si and SiO2/Si Gate Stacks

S.Y. Tan[1]
[1]Department of Electrical Engineering, Chinese Culture University
Taipei, Taiwan 11114, R.O.C.

ABSTRACT

As complementary metal–oxide–semiconductor (CMOS) devices are scaled down into nano-region, SiO_2 dielectric is approaching its physical and electrical limits. Hafnium based oxides are the most promising materials as a replacement for conventional gate dielectrics, due to its much higher dielectric constant (high-k) and stability. The aims of work were to investigate the interface-related issues associating with Hf-based/Si and SiO_2/Si gate stacks. The interfacial properties of dielectric oxide/Si were studied by stacking a different dielectric layer (SiO_2, HfSiO, and HfO_2) on Si substrate. We studied the electrical behavior of dielectric oxide/Si interface by leakage current density-voltage (J-V) and capacitance-voltage (C-V) measurement techniques. The effects of the post-deposition annealing (PDA) treatment on the interface charges of dielectric oxides were presented. We found that the PDA can effectively reduce trapping density and the leakage current, and eliminate hysteresis in the C–V curves. The X-ray photoelectron spectroscopy (XPS) was applied for studying the surface chemical bonding energy at different gate stack structures. The XPS analysis provides a better interpretation of the electrical property. As results, HfSiO films have exhibited a superior performance in terms of thermal stability and electrical characteristics.

INTRODUCTION

Numerous dielectrics with a higher dielectric constant than SiO_2, such as, Y_2O_3, Al_2O_3, La_2O_3, HfO_2, and their compounds have been studied as alternatives for SiO_2 in downscaling of CMOS field effect transistor dimensions [1,2]. Among those potential high-k dielectrics, hafnium based oxides, having potential to form a silicon oxide comparable interface with the bulk Si, are considered as the promising materials for replacing the SiO_2 [3,4]. However, there are some problems to deteriorate its electrical characteristics such as the formation of Hf–silicate layer at the interface between HfO_2 layer and Si substrate, and the steep increase of the interfacial layer due to the crystallization in HfO_2 layer after annealing treatment [2,5]. In order to fabricate high-quality gate insulators, it is necessary to find out the change of the electrical properties according to the different chemical states after annealing treatment. Some groups suggest that the oxide capacitance in accumulation region decreases due to the relative increase of the interfacial layer or increases due to the densification of HfO_2 layer as the annealing temperature increases in the different annealing atmospheres [2,5]. In addition, HfSiO film has been extensively studied recently, mainly because that the silicate-Si is chemically similar to the SiO_2-Si interface, HfSiO can yield lower leakage current than HfO_2 with the same equivalent oxide thickness.

EXPERIMENT

P type (100) oriented silicon substrates with resistivity of 15-25 Ω-cm were used to obtain three different gate structures, SiO_2/Si, $HfSiO/Si$ and HfO_2/Si stacks. For gate dielectric fabrications, SiO_2 layers were thermally grown by dry oxidation (SiO_2) of 2-5 nm using a Vertical Furnace system at temperature around 800 °C. HfO_2 and $HfSiO$ with thickness of 5 nm were deposited by metal organic chemical vapor deposition (MOCVD) at temperature of 500 °C with O_2 gas and deposited rate at 0.6 nm/min. In order to study the thermal stability and crystal structure of Hf-based/Si interface, the post-deposition annealing (PDA) treatment at different temperatures (550 °C and 850 °C) was performed. The films composition and chemical states were examined by X-ray photoelectron spectroscopy (XPS). For the XPS measurements, Al Kα excitation and a VG Scientific Microlab 310F electron analyzer were used. The base pressure was ~10$-$9 Torr for the XPS measurements. The interfacial properties of dielectric oxide/Si were studied by stacking a different dielectric layer (SiO_2, HfSiO, and HfO_2) on Si substrate. We studied the electrical behavior of dielectric oxide/Si interface by leakage current density-voltage (J-V) and capacitance-voltage (C-V) measurement techniques. The leakage current density-voltage (J-V) and capacitance-voltage (C-V) of MIS capacitors were obtained by using Al/Hf-based/Si and Al/SiO$_2$/Si structures. The aluminum film (~5000 Å) was sputtered on top of the wafers and patterned as the top electrode, and another 5000 Å aluminum was also sputtered on the back side to serve as the ohmic contact. The MIS devices and high-frequency (1 MHz) C-V were measured using a Agilent4155 precision meter in a dark chamber at room temperature and the J-V was measured by a Keithley 590 source measure unit.

DISCUSSION

In figure 1 for HfO_2 films with PDA temperatures at 550 °C and 850 °C, the XRD peaks corresponding to polycrystalline Hf are present. Above 850 °C, relatively strong Hf peaks, are observed. For HfSiO films with PDA temperatures at 550 °C and 850 °C (figure 2), the XRD peaks are not observed after a PDA temperature at 850 °C. This indicates that the HfO_2 films became highly textured at a high PDA temperature, whereas, HfSiO films were found to be stable in a high temperature. As the annealing temperature is increasing from 550 °C to 850 °C, the HfSiO film remained as amorphous which indicate that HfSiO is more stable at these temperatures. Therefore, the nucleation of Hf occurs at a higher temperature for HfSiO films than for HfO_2 films.

To investigate the properties at the interface between Hf-based film and Si, the XRD spectra were also supported by XPS results. For HfO_2 in figure 3, the XPS spectra obtained from Hf4f core-levels with PDA at 550 °C and 850 °C. The Hf4f peaks (figure 3) show the difference with PDA temperature. All Hf4f XPS spectra show double peaks, which arise from the spin-orbit splitting of the Hf4f peaks into Hf4f$_{5/2}$ and Hf4f$_{7/2}$. The peaks in samples with PDA at 850 °C are found to have a larger binding energy (BE) than the samples with PDA at 550 °C. This implies that the shift of Hf4f peaks to higher binding energy relative to polycrystalline Hf due to the formation of Hf-OSi bonding at the interface of HfO_2 and bulk Si. This indicates that the low-k interfacial layer between HfO_2 and Si substrate was formed. But, for HfSiO (figure 4), the Hf4f peaks exhibit no difference with PDA temperature. In figure 5 for HfO_2, the XPS spectra of O 1s were composed of two peaks originating from Hf-O and Si-O bonds for PDA at 550 °C and PDA at 850 °C. It shows a shift to higher BE with 850 °C PDA treatment. The XPS results correspond well with XRD analysis due to a high temperature PDA could possibly strengthen the interface layer of Si-O bonds. In figure 6 for HfSiO, the XPS spectra show that the binding energy of Si-O bonds is at about 532.5 eV. The O1s peaks exhibit no difference with PDA temperatures. It seems that the HfSiO films can prevention interfacial layer include effectively.

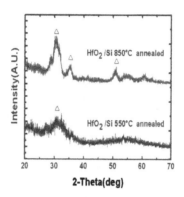

Figure 1. XRD spectra of HfO$_2$/Si stacks with a post deposition annealed temperature (PDA) at 550 °C and 850 °C.

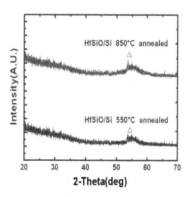

Figure 2. XRD spectra of HfSiO/Si stacks with post deposition annealed temperature (PDA) at 550 °C and 850 °C.

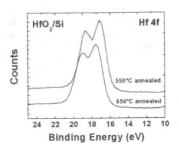

Figure 3. Hf 4f core-level XPS spectra for HfO_2/Si stack with PDA at 550 °C and 850 °C.

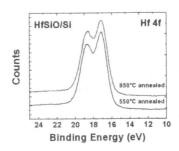

Figure 4. Hf 4f core-level XPS spectra for HfSiO/Si stack with PDA at 550 °C and 850 °C.

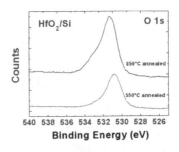

Figure 5. O1s XPS spectra for HfO_2/Si gate stack with PDA at 550°C and 850°C.

18

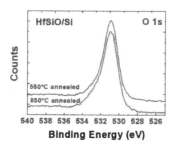

Figure 6. O1s XPS spectra for HfSiO/Si gate stack with PDA at 550°C and 850°C.

Figure 7 and 8 show the C-V curves of HfO_2/Si capacitors resulting from the PDA temperatures at 550 °C and 850 °C. For PDA at 550 °C (see figure 7), there are larger stretch-out near inversion regions compared to PDA at 850 °C (see figure 8). The large hysteresis for the HfO_2 with PDA at 550 °C indicates that the interface trap density was relatively higher. The interface trapping may be due to the structural defects or some sort of deep traps in the as-deposited HfO_2 films. The PDA has found substantially removed defects and the interface trap (detrapping). Apparently the higher the annealing temperature, the larger amounts of charge trapping and detrapping can be expected. The figure 9 shows the for HfSiO with PDA at 850 °C, the C–V curves for the sweeps from accumulation to inversion and back to accumulation are almost identical, and there are no stretch-out in inversion regions. We found that the HfSiO film as a gate dielectric material has demonstrated a good thermal stability and less structural defects, which is compatibly with silicon dioxide (SiO_2) film (figure 10).

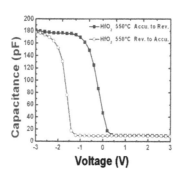

Figure 7. C-V characteristics of Al/HfO$_2$/Si MIS capacitors with PDA temperature at 550 °C.

19

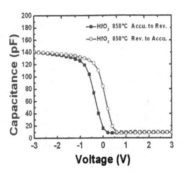

Figure 8. C-V characteristics of Al/HfO$_2$/Si MIS capacitors with PDA temperature at 850 °C.

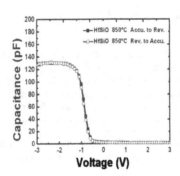

Figure 9. C-V characteristics of Al/HfSiO/Si MIS capacitors with PDA temperature at 850 °C.

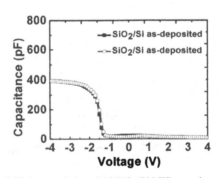

Figure 10. C-V characteristics of Al/SiO$_2$/Si MIS capacitors

The leakage current density of HfSiO/Si and HfO$_2$/Si gate stacks was characterized and shown in figure 11 and figure 12. The average leakage current density of HfSiO samples is found to be much lower than that of HfO$_2$ samples. It is also noticed that the leakage current can be improved by a PDA treatment (~850 °C). Since the high-k film can produce interfacial layer (IL) between the film and substrate, this IL retards the charges transport path along within the film. The leakage current density of SiO$_2$/Si gate stack shows a relatively low leakage and no difference with PDA treatment (figure 13). As results, the post deposition annealing treatment and amorphous texture could effectively reduce bulk and interface trapping density. As a gate dielectric, amorphous films are desirable because of poly-crystalline film can induce fast impurity diffusion along with gain boundaries. Unfortunately, most single oxide high-k dielectrics except Al$_2$O$_3$ have been reported to crystallize at low temperature (< 600 °C). The high-k material such as HfO$_2$ has been reported to be vulnerable to the diffusion of oxygen which causes formation of interfacial layer [6].

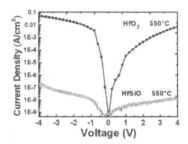

Figure 11. Leakage current density characteristics (J-V) of MIS capacitors with PDA at 550 °C. for HfO$_2$/Si and HfSiO/Si.

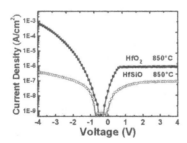

Figure 12. Leakage current density characteristics (J-V) of MIS capacitors with PDA at 850 °C for HfO$_2$/Si and HfSiO/Si.

21

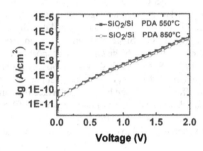

Figure 13. Leakage current density characteristics (J-V) of MIS capacitors with PDA at 550 °C and 850 °C for SiO$_2$/Si.

CONCLUSIONS

The phenomenon of interfacial property between bulk Si and gate dielectric materials (SiO$_2$, HfO$_2$ and HfSiO) was investigated. The XRD results indicated that HfSiO film with PDA at 850 °C was retained as amorphous due to its thermal stability. The XPS was applied for studying the surface chemical bounding energy at different gate stack structures. The results of XPS suggest that the low-k interfacial layer between HfO$_2$ and Si substrate was formed after PDA at 850 °C. However, HfSiO/Si samples exhibited no difference with PDA treatment. It seems that the HfSiO films can prevention interfacial layer include effectively. Thus, it explains the high trap density of HfO$_2$/Si interface observed from C-V and J-V measurements. The PDA treatment can effectively reduced trapping density and the leakage current, and eliminate hysteresis in the C–V curves due to the densification of HfO$_2$. Nevertheless, HfSiO films have exhibited a superior performance on both thermal stability and electrical characteristics.

ACKNOWLEDGMENTS

This work was supported by the National Science Council under award no. NSC97-2221-E-034-001.

REFERENCES

1. M. Houssa, M. Tuominen, M. Naili, V. Afanas'ev, A. Stesmans, S. Haukka, M.M. Heynes, J. Appl. Phys. 87, 8615 (2000).
2. M. Copel, M.A. Gribelyuk, E. Gusev, Appl. Phys. Lett. 76, 436 (2000).
3. N. Zhan, K.L. Ng, M.C. Poon, C.W. Kok, H. Wong, J. Electrochem. Soc. 150, 200 (2003).
4. K.L. Ng, N. Zhan, C.W. Kok, M.C. Poon, H. Wong, Microelectron. Reliab. 43, 1289 (2003).
5. L.F. Schneemeyer, R.B. Van Dover, R.M. Fleming, Appl. Phys. Lett. 75, 1967 (1999).
6. A. Hokazono, K. Ohuchi, M. Takayanagi, Y. Watanabe, S. Magoshi, Y. Kato, T. Shimizu, et al., IEDM Tech. Dig., 639 (2002).

Mater. Res. Soc. Symp. Proc. Vol. 1155 © 2009 Materials Research Society   1155-C03-03

XPS Method to Study the Effect of Heat Treatments and Environment on the Electric Dipole Formation at Metal/High-κ Dielectric Interface

Andrei Zenkevich, Yuri Lebedinskii, Yuri Matveyev and Vladimir Tronin

Moscow Engineering Physics Institute (State University), 115409 Moscow, Russia

ABSTRACT

X-ray photoelectron spectroscopy (XPS) technique is employed *in situ* to quantify changes in the electric dipole layer formed at the metal/dielectric interface. The proposed method is valid in the particular case of discontinuous metal overlayer in contact with dielectric, and allows one to model metal gate effective work function evolution of metal-oxide-semiconductor (MOS) stack following its treatments in different environments. The obtained results on Au / dielectric (dielectric=HfO_2, $LaAlO_3$) corroborate the model that the oxygen vacancies generated in dielectric contribute to the effective work function changes.

INTRODUCTION

The scaling of gate length and gate oxide thickness in complementary metal-oxide-semiconductor (CMOS) transistors has led to the need for the replacement of SiO_2 as a gate dielectric for a novel high-k material [1]. It has also aggravated the problems of poly-Si gate depletion, high gate resistance and dopant penetration from the gate into the channel. In addition, high-k and poly-Si gate often appear incompatible due to Fermi level pinning at the high-k / poly-Si interface which causes high threshold voltages in transistors [2]. As a result, there is an immense interest in metal gate technology. A metal gate material eliminates the gate depletion and boron penetration problems, and also greatly reduces the gate sheet resistance. Metal gate approach requires a dual-work-function metal gate technology for advanced high-performance CMOS devices, analogous to n+ and p+ doped poly-Si gates. Therefore, a major challenge is to find two metals with work functions (WFs) corresponding to the conduction and valence band edges of Si and to integrate them into a CMOS technology.

In the selection of metal gate materials for advanced CMOS applications, an assumption that the work function of a metal on a gate dielectric is the same as that in vacuum has been experimentally observed to be incorrect. It was earlier proposed that metal in contact with dielectric produces so called metal-induced gap states in the forbidden gap of dielectric resulting in charge transfer across the interface [3]. Thus created dipole drives band alignment so that the effective WF of metal in contact with particular dielectric differs from vacuum WF values. Therefore, the effective WF depends on the defects at the interface which can result from heat treatments of MOS stack in different environments.

For the particular case of p-channel field-effect transistor (p-FET) with HfO_2 gate dielectric it was found that noble metals (with high WF) deposited at room temperature with no annealing had no or little dipole formed at the interface (unpinned or weakly pinned Fermi level, EF) [4,5]. However, these metals show instability when annealed above T=500°C in oxygen-deficient conditions, and their EF moves toward midgap, presumably due to oxygen vacancies generated in HfO_2, the latter suggestion confirmed by recent calculations [6].

In this work, we applied the modified version of the earlier proposed x-ray photoelectron spectroscopy (XPS) based method [7] to measure the changes of electrical dipole built up at the metal/high-k dielectric interface (contributing to the "effective WF") following different treatments of the metal-oxide-semiconductor stack. The method is based on the deposition of an ultrathin (1–5 nm) <u>discontinuous</u> island-like metal layer on high-k dielectric surface and *in situ* measurement by XPS of the dielectric constituents core level line shifts <u>with respect</u> to the metal ones. In case there are no chemical reactions at the upper interface (which is the case for noble metal electrodes), these shifts should follow the formation and evolution of the electrical dipole at the metal/high-k dielectric interface (Figure 1).

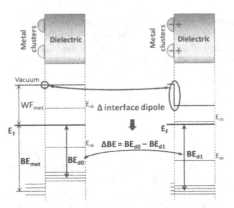

Figure 1. The scheme illustrating the correlation between the dielectric core level binding energy shifts and the changes of electric dipole layer formed at the metal/dielectric interface.

Generally, the electric dipole formed by the array of metallic nanoclusters (ncs) on the dielectric surface induces different potential distribution in dielectric compared to the continuous metal cap. To evaluate the difference and hence the applicability of the proposed XPS method to quantify electric dipole changes, we modeled the potential distribution in case of electric dipole layer formed under metal ncs by numerically solving Laplace equation using the image method. The results of such modeling (to be published elsewhere) performed for the array of metal ncs at our experimental conditions (see Figure 2,a; metal ncs 5-10 nm in size, 5-10 nm separations, the nominal coverage in a nm range, κ=25, the electric dipole ΔU=0.5 V) show that the potential distribution is the same as that of a continuous metal overlayer within the accuracy of our experiment (~0.05 V). The latter conclusion means that XPS can be used to probe the metal/dielectric interface electric dipole layer in case of island-like metal growth (when the continuous metal layer is usually much thicker than ~5 nm XPS probe depth). At the same time, the array of metal ncs at the surface makes the metal/dielectric interface sensitive to different treatments through the uncapped dielectric surface and thus the proposed method provides an opportunity to investigate the effect of these treatments on the electric dipole evolution at the interface.

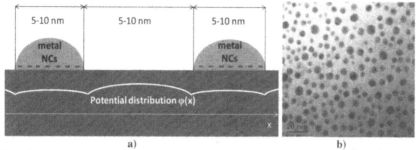

Figure 2. a) the schematic cross-section of metal/dielectric interface used to model the electric potential distribution in dielectric; b) TEM image of ncs array formed on HfO_2 surface following the deposit of Au at 1.5 nm nominal covergae.

EXPERIMENT

The experiments were performed in XSAM-800 spectrometer (Kratos) equipped with two UHV chambers. The preparation chamber (with base pressure of $P\sim10^{-7}$ Pa) was used to controllably deposit Au and Ni at 1-2 nm coverage with submonolayer precision by pulsed laser deposition (PLD) technique. The PLD depositions were performed using YAG:Nd laser ($\lambda=1064$ nm), operating in the Q-switched regime ($\tau=15$ ns) with the variable output energy $E=50\div150$ mJ and the pulse repetition rate $\nu=30$ Hz. The deposition rates of $\sim0.01\div0.1$ monolayer per pulse were initially calibrated with *ex situ* Rutherford backscattering spectrometry (RBS). HfO_2 (3-10 nm) and $LaAlO_3$ (20 nm) high-*k* dielectric layers were grown *ex-situ* by Atomic Layer Deposition (ALD) on chemically oxidized (0.5 nm) and H-terminated Si(100), respectively, and used as a substrate. Prior to metal depositions, the samples were annealed at T=300°C, 3 min. in UHV to remove surface hydrocarbon contaminations, as confirmed by XPS. Upon completion of metal deposition step, the sample can be transferred without breaking vacuum to the analysis chamber equipped with XPS (base pressure $\sim5\cdot10^{-8}$ Pa) and/or resistively heated up to T=500°C, 5 min. either in UHV or in oxygen (P~1 Pa) atmosphere (alternatively, exposed to air) thus allowing to study the effect of the environment on the electrical properties of metal/dielectric interface.

XPS (Mg K_α source, E=1253.6 eV) spectra of Hf4*f*, O1*s*, La4*d*, Al2*p*, Au4*f*, Ni2*p* lines were acquired prior to and after deposition of Au, Ni on HfO_2 and $LaAlO_3$, respectively, as well as upon further annealing steps. Au4*f* $_{7/2}$ line was set at the binding energy $BE_{Au4f}=84$ eV, while Ni2*p* line- at $BE_{Ni2p}=852.5$ eV.

To verify XPS results, metal-insulator-semiconductor (MIS) capacitors were fabricated by PLD deposition of Ni (~50 nm thick) through a shadow mask (gate area: 8×10^{-4} cm^{-2}) on HfO_2(10 nm)/Si samples subjected to *in situ* vacuum annealing (T=500°C) and further air exposure in comparison with the as grown one in comparison to those subjected to vacuum annealing (T=500°C) and air exposure. In-Ga was used as back ohmic contact. The capacitance-voltage (C-V) characteristics were acquired at 100 kHz frequency using Keithley 590 CV Analyzer. The possible correlation between the flatband voltage shift in the C-V curves of

Ni/HfO$_2$/Si stack and the XPS shifts due to variation of electric dipole layer formation at the upper interface, was evaluated as a function of post deposition thermal treatments.

DISCUSSION

The proposed XPS method has been first tested on Au/HfO$_2$ system. Au deposit at 1.5 nm nominal thickness (calibrated with RBS) is condensed in metal ncs covering ~20% of the pre-annealed HfO$_2$ surface area as estimated from XPS signal attenuation curves, and further confirmed *ex situ* by plain-view transmission electron microscopy analysis of the sample (Figure 2,b). The evolution of both Hf4f and O1s lines with respect to Au4f upon Au deposition and further annealing in UHV vs. O$_2$ is given in Fig.3. Since both oxide lines evolve coherently, we conclude HfO$_2$ surface layer undergoes no chemical changes. The exposure of as grown nc-Au/HfO$_2$/Si sample to air results in a small shift of Hf4f and O1s lines binding energy ΔBE= –0.2 eV. Further UHV annealing at T=500°C gives rise to the reverse shift of Hf4f and O1s lines ΔBE=+0.6 eV and +0.7 eV, respectively, while the annealing in O$_2$ at the same T drives the relative binding energy ΔBE=–0.5 eV for both lines back to the position in as grown sample within experimental error of 0.1 eV. The evolution of BE shifts following sample treatments reflects the changes in the value of electrical dipole built-up at nc-Au/HfO$_2$ interface. In our experiment, HfO$_2$ core level lines binding energy are measured with respect to the deposited metal (Au) E_F, and, according to the diagram in Figure 1, the positive ΔBE changes following vacuum annealing indicate that the metal effective work function is driven to the lower values towards Si midgap. On the contrary, the annealing of nc-Au/HfO$_2$/Si stack in O$_2$ (as well as its exposure to the atmospheric air following low temperature UHV pre-annealing) moves the position of E_F down, the cyclic treatments resulting in the corresponding changes in interface dipole (Figure 3). In addition, it is worth noting that for 3 nm thick HfO$_2$ layer on Si (when XPS is also probing the lower HfO$_2$/SiO$_2$/Si interface), small increase of SiO$_2$ interlayer thickness upon each annealing step is noticeable from Si2p line (not shown). The latter effect corroborates the model that oxygen generated in HfO$_2$ during annealing is consumed by Si at the lower interface [6].

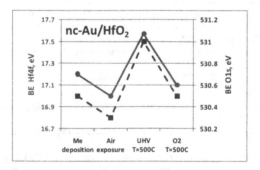

Figure 3. Evolution of Hf4f (solid), O1s (dashed) core level spectra following treatments of Au/HfO$_2$(/Si) stack in different environments.

Similar experiments carried out with nc-Au/LaAlO₃/Si stack yield qualitatively similar results (Figure 4). In particular, reversible and simultaneous changes of La4d and Al2p lines positions $\Delta BE=\pm0.5$ eV following annealing in UHV vs. O_2 are observed. In addition, we attempted to measure ΔBE shifts as a function of temperature *during* UHV annealing, and small positive ΔBE response was indeed observed.

The obtained results indicate that there might be a common mechanism responsible for the evolution of the interface dipole following MOS stack annealing in different environments. It has been suggested [8] that particularly for HfO₂/Si system the effective work function and the flatband voltage of a MOS device with the high work function metal gate (Re, Pt,...) is strongly modulated by the oxygen vacancy concentration in HfO₂ which, in turn, is controlled by the annealing in reducing vs. oxygen atmosphere. In the theoretical work [6] the mechanism explaining the oxygen vacancy generation, at large concentration in ultrathin HfO₂ layers in contact with SiO₂/Si, is proposed and subsequently the Fermi level pinning of the high work function metal is considered in terms of oxygen vacancies in HfO₂. According to the model, the O vacancy generation in HfO₂ is assisted by Si oxidation at the lower interface. In fact, we observe small increase of Si⁴⁺ peak area in Si2p core level line following vacuum annealing of HfO₂(3 nm)/SiO₂(0.5 nm)/Si samples without any metal on top. Similar effects of the annealing environment can be expected also for LaAlO₃ based MOS stacks with high work function metal gates, judging from recent calculations of the oxygen vacancy energy levels in the gap of LaAlO₃ [9]. Our data for Au/HfO₂ as well as for Au/ LaAlO₃ systems corroborate the experimental results [8] as well as support the model [6, 9]. For all investigated systems the vacancies can be removed and subsequently the effective work function of the metal recovered upon annealing in O₂ at T=400-500°C or even upon exposure of the stack with "transparent" metal electrode to atmospheric air. The obtained results underline the importance of the deposited high-k dielectric thermal treatments atmosphere prior to MOS metallization and indicate that low temperature annealing can affect the value of electrical dipole at the metal/dielectric interface through the oxygen vacancy generation in dielectric.

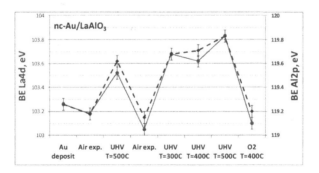

Figure 4. Evolution of La4d (solid), Al2p(dashed) core level spectra following/during treatments of Au/LaAlO₃(/Si) stack in different environments.

The XPS based method is proposed to characterize the effect of heat treatments on the value of electric dipole formed at metal gate/high-k dielectric interface in MOS stacks. It was applied to investigate Au/HfO_2 and $Au/LaAlO_3$ interface properties following heat treatments in vacuum vs. in O_2 or air exposure. The obtained results corroborate the theoretical model ([6]) which suggests that oxygen vacancies generated in dielectrics affect the magnitude of electric dipole built at metal/dielectric interface.

ACKNOWLEDGEMENTS

The authors would like to thank L. Lamanga of National Laboratory CNR-INFM-MDM (Italy) for ALD growth of dielectric layers, P. Chernykh of Skobeltsyn Insitute of Nuclear Physics (Moscow) for RBS analysis and A. Orekhov of Institute of Crystallography (Moscow) for TEM analysis.

REFERENCES

1. G.D. Wilk, R.M. Wallace and J.A. Anthony, J. Appl. Phys. 89, 5243-5275 (2001).
2. C. Hobbs, L. Fonseca, V. Dhandapani, S. Samavedam, B. Taylor, J. Grant, L. Dip, D. Triyoso, R. Hegde, D. Gilmer, R. Garcia, D. Roan, L. Lovejoy, R. Rai, L. Hebert, H. Tseng, B. White and P. Tobin, Symposium on VLSI Technology Digest, 9-10 (2003).
3. Yee-Chia Yeo, P. Ranade, Tsu-Jae King and Chenming Hu, IEEE Electron. Dev. Lett. 23, 342-344 (2002).
4. M. Koyama, Y. Kamimuta, T. Ino, A. Kancko, S. Inumiya, K. Eguchi, M. Takayanagi, and A. Nishiyama and Tech. Dig., Int. Electron Devices Meet., 499 (2004).
5. V. Afanasev, M. Houssa, A. Stesmans, and M. M. Heyns, J. Appl. Phys. 91, 3079 (2002).
6. J. Robertson, O. Sharia and A. A. Demkov, App. Phys. Lett. 91, 132912 (2007).
7. Yu. Lebedinskii, A. Zenkevich and E.P. Gusev, J. Appl. Phys. 101, 074504 (2007).
8. E. Cartier, F. R. McFeely, V. Narayanan, P. Jamison*, B. P. Linder, M. Copel, V. K. Paruchuri, V.S. Basker, R. Haight, D. Lim, R. Carruthers, T. Shaw, M. Steen, J. Sleight, J. Rubino, H. Deligianni, S. Guha, R. Jammy and G. Shahidi, Symposium on VLSI Technology Digest, 230 (2005).
9. K. Xiong, J. Robertson and S.J. Clark, Microelectron. Eng. 85, 65–69 (2008).

Mater. Res. Soc. Symp. Proc. Vol. 1155 © 2009 Materials Research Society   1155-C08-05

Synchrotron Radiation and Conventional X-ray Source Photoemission Studies of γ-Al₂O₃ Thin Films Grown on Si(111) and Si(001) Substrates by Molecular Beam Epitaxy

M. El Kazzi[1], C. Merckling[2], G. Grenet[3], G. Saint-Girons[3], M. Silly[1], F. Sirotti[1], G. Hollinger[3]

[1] Synchrotron SOLEIL, L'Orme des Merisiers, 91192 Gif Sur Yvette Cedex - France
[2] IMEC, Kapeldreef 75, B-3001 Leuven, Belgium
[3] INL (UMR 5270), Ecole Centrale de Lyon, 69134 Ecully Cedex - France

ABSTRACT

High-resolution synchrotron radiation X-ray photoelectron spectroscopy (HRXPS) is used to study the chemical bonding at the Al₂O₃/Si(001) and Al₂O₃/Si(111) interfaces. In both cases, the Si2p spectra recorded at 180 eV photon energy provides evidence a thin interfacial layer rich in Si-O bonding. On the other hand, conventional AlKα X-ray source angular measurements clearly indicate that there are two in-plane orientations for Al₂O₃/Si(111) : [11-2]Al₂O₃(111)//[11-2]Si(111) and [-1-12] Al₂O₃(111)//[11-2]Si(111) but four in-plane orientations for Al₂O₃/Si(001) : [11-2] Al₂O₃(111)//[100]Si(001), [11-2]Al₂O₃(111)//[010]Si(001), [11-2]Al₂O₃(111)//[-100]Si(001), and [11-2]Al₂O₃(111)//[0-10]Si(001).

INTRODUCTION

Growing well-crystallised oxide thin films on silicon is the subject of intense activity due to numerous potential applications in nanotechnologies.[1,2] Among oxides, Al₂O₃ is a very promising material because of its stability on silicon for temperatures and oxygen pressures up to 900°C and 10⁻⁵ Torr, respectively.[3, 4, 5, 6, 7]. If grown as a buffer layer between high-κ oxides and silicon, it can also provide an efficient barrier against silicates and silica formation in CMOS devices. [8, 9] However, the interest of studying and controlling Al₂O₃ growth on silicon goes far beyond because this could make the integration of high performance micro-optoelectronic functionalities on Si wafers possible.

Despite a large lattice mismatch (the lattice parameters for the spinel γ-Al₂O₃ and silicon are a = 0.791 nm and a₀= 0.5431nm respectively), the epitaxy of Al₂O₃ on silicon has already been reported successful by several groups. On Si(111), the growth is that of a (111)-oriented spinel-like γ-Al₂O₃. [3, 4, 6] On Si(001), the growth is more complex. The two first monolayers are (001)-oriented. At this early stage, the RHEED pattern is (1×5) with a four-fold symmetry. After that, a transition occurs and the growth direction changes from [001] to [111]. The surface exhibits a 12-fold symmetry subsequent to a growth with two domains rotated by 90° from each other.[5, 6]

In this work, we have studied the Al₂O₃/Si(111) and Al₂O₃/Si(001) interfaces as a function of photon energy using synchrotron-radiation X-ray photoemission spectroscopy XPS (HRXPS) to take advantage of the high flux to investigate interfacial bonding. The in-plane crystallographic orientation has been studied using conventional source X-ray photoelectron diffraction (XPD) to explore the local crystallographic arrangement.[10]

RESULTS

Al_2O_3 layers were grown on Si(111) and Si(001) in a Riber 2300 MBE reactor equipped with a 30 kV Reflection High Energy Electron Diffraction (RHEED) system. They were prepared by electron gun evaporation of Al_2O_3 single crystal under molecular oxygen with a growth temperature of 850°C and an oxygen partial pressure around 10^{-5} Torr. The Al_2O_3 growth rate was controlled in-situ using a mass spectrometer. On Si(001) The growth was stopped just after the transition from [001] to [111] orientation. This transition corresponds to a layer thickness of 1nm thus permitting the measurement of both the substrate Si2p core level and the Al2p and O1s core levels from the Al_2O_3 layer.

1. HRXPS measurements

HRXPS measurements were performed on the soft-x-ray beamline (TEMPO) at (SOLEIL) synchrotron in France. The end-station is equipped with a high-performance electron analyzer SCIENTA-SES 2002 with an acceptance angle of 1°. Measurements were performed at normal emission. Photon energies of 700 eV and 180 eV were chosen to achieve high volume and interface sensitivity, respectively. At these photon energies, the photon energy resolution is 140 meV and 36 meV while the electron energy resolution is 60 meV and 30 meV, respectively.

Figure 1 compares Al2p and Si2p core level spectra for Al_2O_3/Si(111) (figure 1.(a) (b)) and Al_2O_3/Si(001) (figure 1.(c) (d)) for three photons energies, i.e., 180 eV and 700 eV two from synchrotron radiation and 1486.6 eV from conventional Alkα source. At the two highest photon energies (700eV and 1486.6 eV), both Al2p and Si2p peaks exhibit a sharp structure at 75 eV and 99.7 eV respectively, without any secondary features at lower or higher binding energy. This constitutes a good indication that the interface between the Al_2O_3 film and the Si substrate is abrupt without any silicate or silica even with high temperature growth and oxygen pressure. Moreover, the alumina film is completely oxidized without alumina sub-oxide (Al_2O_x).

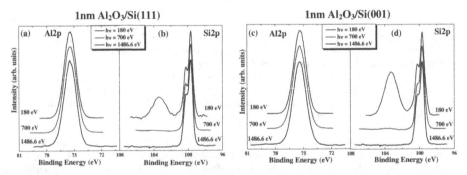

Figure 1: Comparison of the Al2p and Si2p core levels spectra for 1nm-thick Al_2O_3 layer on Si(111) and Si(001) versus photon energy 180 eV, 700 eV (synchrotron radiation) and 1486.6 eV (conventional Alkα source).

Let us turn now to the spectra at the lowest photon energy (180 eV), which is more sensitive to interfacial behavior. The Al2p core level in figure 1.(a) and (c) remains unchanged but the Si2p core level in figure 1.(b) and (d) presents two peaks, a narrow peak at 99.7 eV and a broad peak at higher binding energy, which can be attributed to chemically shifted interface oxidation chemical states. This peak is more intense for $Al_2O_3/Si(100)$ than for $Al_2O_3/Si(111)$ and exhibits a small difference in binding energy shift, i.e., 3.5eV and 3.3eV, respectively. This position is close to that observed for Si^{4+} found at 3.9 eV [11], 3.84 eV [12], 3.64 [13], 3.54 eV [14] for SiO_2. If it were due to SiO_2, the thickness of the interfacial SiO_2 layer would be 0.25 nm and 0.5 nm respectively to take into account the difference in intensity. Another origin could be the Si–O–Al bonds connecting Al_2O_3 and Si lattices via a sub-lattice of oxygen. In this case, the peak intensity depends on the number of Si–O–Al bonds and therefore on the substrate orientation.

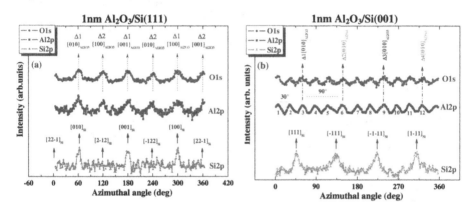

Figure 2: XPD azimuth curves of O1s, Al2p, Si2p core levels recorded for polar angle ($\theta = 55°$) for 1nm-thick Al_2O_3 layer on (a) Si(111) and (b) Si(001).

2. Conventional source AlKα angular measurements

The analysis was performed in a Vacuum Science Workshop (VSW) spectrometer equipped with a focused unpolarized monochromatic AlK_α X-ray (hv = 1486.6 eV) source and a hemispherical analyzer with an acceptance angle around 3°. The XPD curves were obtained by rotating the sample via step-by-step motors with a precision better than 1 deg. The angle between the x-ray source and the analyser is fixed at 50 deg.

Figure 2 shows azimuth curves at a polar angle of 55° for Al2p (kinetic energy = 1411 eV), O1s (kinetic energy = 956 eV) and Si2p (kinetic energy = 1386 eV) for a 1nm-thick Al_2O_3 layer on Si(111) (figure 2.(a)) and on Si(001) (figure 2.(b)).

In figure 2a, Si2p azimuth curves exhibit a clear three-fold symmetry typical of a Si(111) face with intense peaks every 120°, i.e., towards the Si <001> directions ($\varphi = 60°, 180°, 300°$) in agreement with the literature.[15, 16, 17] A three-fold symmetry is also expected for Al2p and O1s in $Al_2O_3(111)$. In fact, the symmetry is six-fold because of the presence of two in-plan orientations labelled Δ1 and Δ2 rotated of 180° from each other.[10] If the Al2p and O1s <010> XPD peaks for Δ1 are lined up with the Si2p <010> XPD peaks, the Al2p and O1s XPD <010> peaks for Δ2 corresponds to Si2p <-122> directions for which only a weak XPD peak is observed. In other words, the in-plane relationships for the two domains Δ1 and Δ2 are defined by the two orientations «direct» [11-2]Al$_2$O$_3$(111)//[11-2]Si(111) and «mirror» [-1-12] Al$_2$O$_3$(111)//[11-2]Si(111) respectively.

In Figure 2b, Si2p azimuth curves presents a four-fold symmetry typical of (001)face with intense peaks every 90°, i.e., towards the Si <111> directions (φ = 45°, 135°, 225°, 315°) in good agreement with the literature.[18] As the Al$_2$O$_3$ layer thickness is just above the [100] to [111] orientation transition, the expected symmetry is expected a multiple of 3. In fact, the Al2p and O1s azimuth curves are 12-fold with intense peaks every 30° due to four Al$_2$O$_3$(111) in-plan orientations labelled Δ1, Δ2, Δ3, and Δ4 rotated by 90° from each other. The Si <111> XPD peaks are lined up with minima of Al2p and O1s azimuth curves. These minima are at 15° from the Al$_2$O$_3$ <010> XPD peaks. It results four in-plane relationships which are Δ1 [11-2]Al$_2$O$_3$(111)//[100]Si(001), Δ2 [11-2]Al$_2$O$_3$(111)//[010]Si(001), Δ3 [11-2]Al$_2$O$_3$(111)//[-100]Si(001), Δ4 [11-2]Al$_2$O$_3$(111)//[0-10]Si(001).

CONCLUSION

In this work, we have studied the interface between alumina and silicon versus substrate orientation. Synchrotron radiation is used to highlight the interface chemical bonding. At low photon energy, a XPS feature in Si2p core level spectrum arises at around 3.5eV from the main peak. From a XPS point of view, this feature can be due either to an ultra thin SiO$_2$ undetectable at higher photon energy or to a Si-O-Al bonding. The greater intensity observed for a Si(001) substrate than for a Si(111) substrate can be the result of a higher surface reactivity as well as a higher number of interface Si-O-Al bonds due to a more complex interface. Conventional source angular measurements indicates that the connection is simpler between Al$_2$O$_3$ and Si(111) than between Al$_2$O$_3$ and Si(001). For Al$_2$O$_3$/Si(111), there are only two in-plane orientations : a «direct» relationship, i.e., [11-2]Al$_2$O$_3$(111)//[11-2]Si(111) and a «mirror» relationship, i.e., [-1-12] Al$_2$O$_3$(111)//[11-2]Si(111) while for Al$_2$O$_3$/Si(001), there are four in-plane orientations [11-2]Al$_2$O$_3$(111)//[100]Si(001), [11-2]Al$_2$O$_3$(111)//[010]Si(001), [11-2]Al$_2$O$_3$(111)//[-100]Si(001), [11-2]Al$_2$O$_3$(111)//[0-10]Si(001). This photoemission study must be completed by vibrational techniques such as polarization-modulation infrared reflection absorption spectroscopy (PM-IRAS) by which the interfacial bonding can be experienced more directly.

ACKNOWLEDGMENTS

The authors would like to thank P. Regreny, N. Bergeard for helpful discussions and J. B. Goure, C. Botella, C. Chauvet for technical assistance.

32

REFERENCES

1. D. G. Schlom, J. H. Haeni, J. Vac. Sci. Technol. A **20**, 1332 (2002).
2. J.-P. Locquet, C. Marchiori, M. Sousa, J. Fompeyrine and J. W. Seo, J. Appl. Phys. **100**, 051610 (2006).
3. T. Okada, M. Ito, K. Sawada, M. Ishida, J. Crys. Growth, **290**, 91 (2006).
4. S. Y. Wu, M. Hong, A. R. Kortan, J. Kwo, J. P. Mannaerts, W. C. Lee, Y. L. Huang, Appl. Phys. Lett. **87**, 091908 (2005).
5. C. Merckling, M. El Kazzi, G. Delhaye, M. Gendry, G. Saint-Girons, G. Hollinger, L. Largeau, G. Patriarche, Appl. Phys. Lett. **89**, 232907 (2006).
6. C. Merckling, M. El Kazzi, G. Saint-Girons, V. Favre-Nicolin, L. Largeau, G. Patriarche, V. Favre-Nicolin, O. Marty, G. Hollinger, J. of Appl. Phys. **102**, 024101 (2007).
7. C. Merckling, M. El Kazzi, V. Favre-nicolin, M. Gendry, Y. Robach, G. Grenet, G. Hollinger, Thin Solid Films **515**, 6479 (2007).
8. L. Beccerra, C. Merckling, M. El Kazzi, N. Baboux, B. Vilquin, G. Saint-Girons, C. Plossu, G. Hollinger, J. Vac. Sci. Technol. B **27**, 384 (2009).
9. C. Merckling, M. El Kazzi, L. Beccera, L. Largeau, G. Patriarche, G. Saint-Girons, G. Hollinger, Microelectronic Engineering, **84**, 2243-2246 (2007).
10. M. El Kazzi, G. Grenet, C. Merckling, G. Saint-Girons, C. Botella, O. Marty, G. Hollinger, accepted in Phys. Rev. B.
11. F.J Himpsel, F.R. McFeely, A. Taleb-Ibrahimi, J. A. Yarmoff, G. Hollinger, Phys. Rev. B **38**, 6084 (1988).
12. Y. Enta, Y Miyanishi, H. Irimachi, M. Niwano, M. Suemitsu, N. Miyamoto, E. Shigemasa, H. Kato, Phys. Rev. B **57**, 6294 (1998).
13. J. H. Oh, H. W. Yeom, Y. Hagimoto, K. Ono, M. Oshima, N. Hirashita, M. Nywa, A. Toriumi, A. Kakizaki, Phys. Rev. B **63** 205310 (2001).
14. S. Dreiner, M. Schürmann, C. Westphal, H. Zacharias, Phys. Rev. Lett. **86**, 4068 (2001).
15. S. Van, D. Steinmetz, D. Bolmont, J. J. Koulmann, Phys. Rev. B **50**, 4424 (1994).
16. L. Simon, D. Aubel, L. Kubler, J. L. Bischof, Gewinner, J. Appl. Phys. **81**, 2635 (1997).
17. S. Bengio, M. Martin, J. Avilla, M. C. Asensio, H. Ascolani, Phys. Rev. B **65**, 205326 (2002).
18. S. Dreiner, M. Schürmann, C. Westphal, Phys. Rev. Lett. **93**, 126101 (2004).

33

Mater. Res. Soc. Symp. Proc. Vol. 1155 © 2009 Materials Research Society   1155-C09-11

Ligand Structure Effect on a Divalent Ruthenium Precursor for MOCVD

Kazuhisa Kawano[1,2], Hiroaki Kosuge[1], Noriaki Oshima[1], Tadashi, Arii[3], Yutaka Sawada[4], and Hiroshi Funakubo[2]
[1]TOSOH Corporation, Hayakawa 2743-1, Ayase, Kanagawa 252-1123, Japan
[2]Department of Innovative and Engineered Materials, Tokyo Institute of Technology, J2-43, 4259 Nagatuta-cho, Midori-ku, Yokohama, Kanagawa 226-8502, Japan
[3]Rigaku Corporation, 3-9-12 Matsubara, Akishima, Tokyo 196-8666, Japan
[4]Tokyo Polytechnic University, 1583 Iiyama, Atsugi, Kanagawa 243-0297, Japan

ABSTRACT

Thermal properties of five divalent ruthenium precursors with three types of structures were examined by thermal analyses. Their volatilities and the relationships between their structure and thermal stability were compared by TG analysis. Precursor volatility was found to be inversely proportional to molecular weight. The DSC result showed that substituting a linear pentadienyl ligand for a cyclopentadienyl ligand decreased the thermal stability of a precursor and precursors could be liquefied by attaching an alkyl group longer than methyl group to a Cp ligand. As a result of TG-MS analyses for Ru(DMPD)(EtCp) and Ru(EtCp)$_2$, 2,4-dimethyl-1,3-pentadiene was found to be a thermolysis product of Ru(DMPD)(EtCp) though no thermolysis products of Ru(EtCp)$_2$ were observed. These results show that the volatility and decomposition temperature of a divalent ruthenium precursor can be designed by adjusting the precursor's structure.

ŕ

INTRODUCTION

Continuing growth of memory capacity for the requirement of miniaturizing semiconductor memory chips will require the exploitation of three-dimensional memory structures even if the capacitor dielectrics having a permittivity higher than that of SiO_2, such as tantalum pentoxide [1] or strontium titanate, [2] are used. High-permittivity dielectrics for next-generation dynamic random access memory devices have therefore been investigated extensively. On the other hand, precious metals such as ruthenium, iridium, and platinum are candidates of electrode materials for these dielectrics, as well as their oxides. Among them, ruthenium was especially promising because of their low resistivity, excellent chemical stability, and good dry-etching properties. [3-6] As the deposition method, metalorganic chemical vapor deposition (MOCVD) has great advantages compared to the methods sputtering with regards to good throughput, composition control, and conformal deposition. [7] Among the variety of ruthenium precursors for MOCVD, the most promising one was bis(ethylcyclopentadienyl)ruthenium [Ru(EtCp)$_2$] because of its high vapor pressure, low melting point, and high stability.[8] A serious problem, however, is a long "incubation time" and therefore then ruthenium seed layer by the sputter deposition is required to diminish the incubation time. [9] Our group has already reported that (2,4-dimethylpentadienyl)(ethylcyclopentadienyl) ruthenium [Ru(DMPD)(EtCp)] showed much shorter incubation time than Ru(EtCp)$_2$ and that metallic ruthenium films could be deposited from Ru(DMPD)(EtCp) without a seed layer. [10] Moreover, conformal deposition go both metallic ruthenium and ruthenium oxide were ascertained from Ru(DMPD)(EtCp) even on a patterned substrate. [11,12] Kim *et al.* also reported conformal deposition of Ru films from Ru(DMPD)(EtCp) on a substrate with an aspect ratio of 17 by atomic layer deposition. [13]

In this study we evaluated five divalent ruthenium precursors as MOCVD precursors for measuring their volatilities and their decomposition characteristics. We also examined the relationships between the thermal stability and ligand structure by Simultaneous thermogravimetry and mass spectrometry (TG-MS) to analyze the decomposition mechanisms of two of the precursors, Ru(EtCp)$_2$ and Ru(DMPD)(EtCp).

EXPERIMENT

Bis(ethylcyclopentadienyl)ruthenium (1) were supplied from Aldrich. Bis(2,4-dimethylpentadienyl)ruthenium (5) was synthesized according to the previous report. [14] The synthetic route of half-open ruthenocene type precursors 2,3 and 4 was shown in Scheme I.

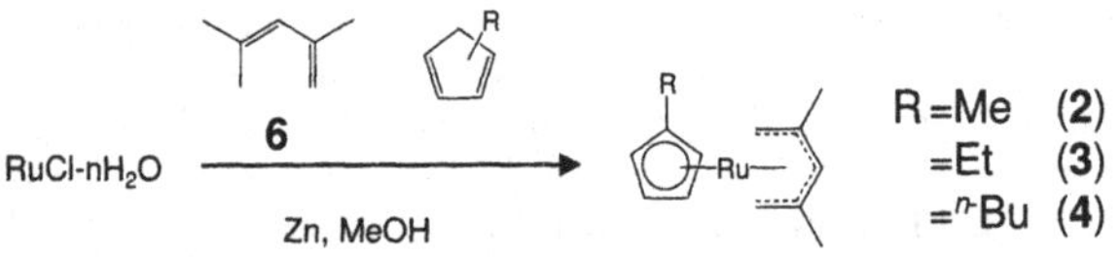

Scheme I. Synthesis of half-open ruthenocene

To a suspension of 8.0 g (122 mmol) of zinc in 2,4-dimethyl-1,3-pentadiene (6, 3.0 g, 31 mmol) and alkylsubstituted cyclopentadiene monomer (2.4 mmol) freshly distilled before use, 20 mL of an ethanol solution of ruthenium trichloride hydrates (0.6 g, 2.3 mmol) was added dropwise over 50 min at 0°C under an argon atmosphere. The mixture was stirred at room temperature for 30 min and then heated to 70°C and stirred for 2 h. After cooling to room temperature, the mixture was filtered to separate a small amount (trace) of precipitate. The filtrate was concentrated *in vacuo* to give a pasty mixture. Pentane (10 mL) was added four times and the paste was dispersed. The supernatant was decanted off, the collected pentane solution was evaporated *in vacuo*, and the residue was purified on an alumina column using hexane to give the target compounds (yields 16-45%). Compounds ware identified with EI-MS, IR, and NMR.

Volatilities of the precursors were estimated by thermogravimetry (TG) analysis under an Ar flow rate of 400 mL/min at a heating rate of 10°C/min using a TG/DTA 200 (Seiko Instruments Inc.). Decomposition characteristics were analyzed using a differential scanning calorimetry (DSC 200, Seiko Instruments Inc.) under a N$_2$ flow of 50 mL/min at a heating rate of 10°C/min. Simultaneous thermogravimetry and mass spectrometry (TG-MS) was performed on a system made by Rigaku Corp. Fig. 1 shows the schematic diagram and setting temperature of the TG-MS apparatus. The exhaust gas from a TG-DTA 8120 (Rigaku Corp.) was led into a mass spectrometer (QP5050A, Shimadzu Corp.), and a 5 cm length of that exhaust line could be heated to temperatures as high as 500°C. The temperature of the other parts of the line was kept at 290°C to avoid condensation of the precursor. Ruthenium precursors vaporized in the TG chamber and could then be heated to 500°C before reaching the MS apparatus, where partially decomposed thermolysis products could be detected.

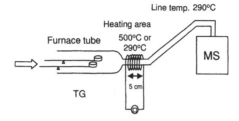

Figure 1. Schematic diagram of TG-MS apparatus

DISCUSSION

The structures and molecular weights of the three types of the organometallic ruthenium precursors examined in this study are shown in Table I.

Table I. Three kinds of ruthenium compounds and their molecular weights

Classification	Ruthenocene type	Half-open ruthenocene type	Bis-open ruthenocene type
Structure			
Molecular weight	**1** Mw 287.37	R =Me (2) Mw 287.37 =Et (3) Mw 289.39 =^{n}Bu (4) Mw 317.44	**5** Mw 291.40

The ruthenocene type [Ru(EtCp)$_2$; **1**] consisted of a ruthenium atom and two ethylcyclopentadienyl groups. Each of the three half-open ruthenocene types consisted of a ruthenium atom, a "linear" pentadienyl group (2,4-dimethylpentadienyl), and a variously substituted cyclopentadienyl group: (2,4-dimethylpentadienyl)(methylcyclopentadienyl) ruthenium; **2**, Ru(DMPD)(EtCp); **3**, and (2,4-diemthylpentadienyl)(n-butylcyclopentadienyl) ruthenium; **4**. The bis-open ruthenocene type consisting of a ruthenium atom and two "linear" pentadienyl groups [bis(2,4-dimethylpentadienyl)ruthenium; **5**] was prepared as described elsewhere. [14] The molecular weights of the precursors are also listed in the Table I for later discussion on the relationship between the volatility and the molecular weight.

The results of TG analyses and relation between molecular weight of precursors and 5% weight loss temperature read from TG are shown in Fig. 2 (a) and (b), respectively. Each of these ruthenium precursors except precursor **5** vaporized completely under the experimental conditions. Precursor **5** started to decompose before vaporizing completely as mentioned below. The volatility of ruthenium precursors shown in Fig. 2 (b) increased monotonically with decreasing molecular weights. These precursors' volatilities were inversely proportional to their molecular weights.

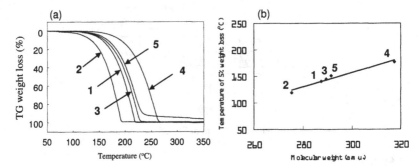

Figure 2. (a); TG charts of five ruthenium precursors. (b); Relation between molecular weights of ruthenium precursors and 5% weight loss temperature in (a).

Fig. 3 shows the thermal decomposition characteristics of the ruthenium precursors analyzed by DSC. Samples were sealed in a stainless vessel under Ar and heated up to 500°C. Exothermic peaks in the chart showed the decomposition temperatures of the each precursor.

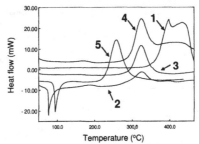

Figure 3. DSC curves of five ruthenium precursors.

Precursor **5**, which had a bis-open ruthenocene structure, started to decompose at 210°C. This is why it did not completely vaporize during the TG analysis shown in Fig. 2(a). Precursors **2, 3,** and **4,** which have half-open ruthenocene structures, started to decompose at almost the same temperature of 270°C, regardless of the length of alkyl group attached to the cyclopentadienyl (Cp) ligand. Precursor **1,** which has the ruthenocene structure, started to decompose at 350°C. This means that the ruthenocene type is the most thermally stable among the three types we examined. These results clearly show that the decomposition characteristics of the precursors depended on the ligand structures. The "linear" pentadienyl ligand makes the decomposition temperature lower than the Cp ligand. It was previously reported that the incubation time could be shorten by changing ruthenium precursor from the precursor **1** to the precursor **3**. [10] This DSC result implies that precursors **2, 4,** and **5** are also potentially useful as MOCVD precursors. The endothermic peaks at 76°C and 94°C in the DSC curves for precursors **2** and **5** are the melting points, which mean these precursors are solids at room temperature. Precursor **3,** on the other hand, has a melting point of 17°C is thus a liquid at room temperature. Precursors can be liquefied by attaching longer alkyl groups to a Cp ligand in a half-open ruthenocene type precursor than methyl group but also decrease its volatility due to increasing its molecular weight. Precursor **4** has a longer alkyl group (butyl group; C4) attached to a Cp ligand than precursor **3** and is also a liquid at room temperature. The lower volatility of precursor **4** than that of precursor **3** is easily understood from the TG results even though the decomposition characteristics of these two precursors are basically the same.

Precursors **1** and **3** were selected as test precursors for TG-MS analyses to reveal the effects of ligands. At first, TG-MS test was carried out under the condition that the 5 cm heating area in Fig .1 was set at 290°C and Fig. 4 was the obtained results of mass spectra.

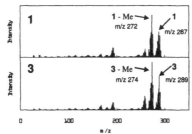

Figure 4. MS charts obtained from TG-MS measurement of **1** and **3** without partially heating in Fig. 1.

Their parent-ion peaks (at m/z 287 for precursor **1** and m/z 289 for precursor **3**) are observed and their fragment-ion peaks due to lost methyl groups (at m/z 272 for precursor **1** and m/z 274 for precursor **3**) are also observed. The results of tests carried out with the heated area of the exhaust line in Fig.1 at 500°C are shown in Fig. 5(a).

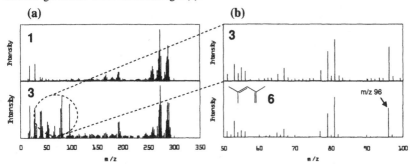

Figure 5. **(a)**: MS charts obtained from TG-MS measurement of **1** and **3** with partially heating at 500°C in Fig.1, **(b)**: Enlarged drawing of low-molecular weight area mass spectrum of **3** in **(a)** and the mass spectrum of 2,4-dimethyl-1,3-pentadiene (**6**)

Significant change at low mass range was observed in the spectrum of the precursor **3**, even though a small change was observed in the spectrum of the precursor **1** compared to the results without partial heating at 500°C. Fig. 5(b) shows enlarged drawing of the low-molecular-weight part of the mass spectrum of precursor **3** in Fig. 5(a) and also shows the mass spectrum of 2.4-dimethyl-1,3-pentadiene (compound **6**). A parent- ion peak of the compound **6** (m/z 96) is observed in the chart of precursor **3,** and other fragmentation peaks are also quite similar to each other. This indicates that precursor **3** partially decomposed at 500°C and formed compound **6,** which is the ligand of precursor **3.** This result supports the hypothesis on decomposition mechanism speculated by Shibutami *et al.* [15] and also be the evidence that a linear pentadienyl ligand was the reason that the decomposition temperature of precursor **3** was lower than that of precursor **1.** These results reveal that the stability and reactivity of the divalent ruthenium precursors can be adjusted by changing the ligand structure of the precursors.

CONCLUSIONS

Three types of divalent ruthenium precursors were analyzed by TG and DSC. The volatilities of the ruthenium precursors could be explained by their molecular weights. Precursors could be liquefied by attaching a long alkyl group but decreased their volatility. The thermal stability of the precursors could be tuned by changing their ligands. A linear pentadienyl ligand can make a divalent ruthenium precursor more thermally unstable than one with a cyclopentadienyl ligand. The effect of adoption of linear pentadienyl ligand to a ruthenium precursor was proved by TG-MS analysis.

REFERENCES

1. T. Aoyama, S. Saida, Y. Okayama, M. Fujisaki, K. Imai, and T. Arikado, *J. Electrochem. Soc.*, **143**, 977 (1996).
2. H. Yamaguchi, P. Y. Lesaicherre, T. Sakuma, Y. Miyasaka, A. Ishitani, and M. Yoshida, *Jpn. J. Appl. Phys.*, **32**, 4069 (1993).
3. P. Y. Lesaicherre, S. Yamamichi, K. Takemura, H. Yamaguchi, K. Tokashiki, Y. Miyasaka, M. Yoshida, and H. Ono, *Integrated Ferroelectrics*, **11**, 81 (1995).
4. W. Pan and S. B. Desu, *J. Vac. Sci. and Technol.*, **B12**, 3208 (1994).
5. D. P. Vijay, S. B. Desu, and W. Pan, *J. Electrochem. Soc.*, **140**, 2635 (1993).
6. S. Saito and K. Kuramasu, *Jpn. J. Appl. Phys.*, **31**, 135 (1992).
7. J. F. Scott, F. D. Morrison, M. Miyake, P. Zubko, Xiaojie Lou, V. M. Kugler, S. Rios, M. Zhang, T. Tatsuta, O. Tsuji, and T. J. Leedham, *J. Amer. Ceramic Soc.*, **88**, 1691 (2005).
8. S. Y. Kang, K. H. Choi, S. H. Lee, C. S. Hwang, and H. J. Kim, *J. Electrochem. Soc.*, **147**, 1161 (2000).
9. M. Kadoshima, T. Nabatame, M. Hiratani, and Y. Nakamura, *Jpn. J. Appl. Phys., Part 2*, **41**, L437 (2002).
10. T. Shibutami, K. Kawano, N. Oshima, S. Yokoyama, and H. Funakubo, *Electrochem. Solid-State Lett.*, **6**, C117 (2003).
11. K. Kawano, A. Nagai, H. Kosuge, T. Shibutami, N. Oshima, and H. Funakubo, *Electrochem. Solid-State Lett.*, **9**, C107 (2006).
12. K. Kawano, H. Kosuge, N. Oshima, and H. Funakubo, *Electrochem. Solid-State Lett.*, **9**, C175 (2006).
13. S. K. Kim, S. Y. Lee, S. W. Lee, G. W. Hwang, C. S. Hwang, J. W. Lee, and J. Jeong, *J. Electrochem. Soc.*, **154**, D95 (2007).
14. C. Brekel, J. Brekel, and J. Bloem, *Philips Res. Rep.*, **32**, 234 (1977).
15. T. Shibutami, K. Kawano, N. Oshima, S. Yokoyama, and H. Funakubo, *Mater. Res. Soc. Symp. Proc.*, **748**, U12.7 (2003).

Mater. Res. Soc. Symp. Proc. Vol. 1155 © 2009 Materials Research Society
1155-C11-06

FTIR Study of Copper Agglomeration During Atomic Layer Deposition of Copper

Min Dai,[1] Jinhee Kwon,[2] Yves J. Chabal,[2] Mathew D. Halls,[3] and Roy G. Gordon[4]
[1]Laboratory for Surface Modification, Rutgers University, 136 Frelinghuysen Road, Piscataway, NJ 08854, U. S. A.
[2]Department of Materials Science and Engineering, The University of Texas at Dallas, Richardson, TX 75080, U.S.A.
[3]Materials Science Division, Accelrys Inc., San Diego, CA 92121, U.S.A.
[4]Department of Chemistry and Chemical Biology, Harvard University, Cambridge, MA 02138, U.S.A.

ABSTRACT

The growth of of metallic copper by atomic layer deposition (ALD) using copper(I) di-sec-butylacetamidinate ($[Cu(^sBu\text{-amd})]_2$) and molecular hydrogen (H_2) on SiO_2/Si surfaces has been studied. The mechanisms for the initial surface reaction and chemical bonding evolutions with each ALD cycle are inferred from in situ Fourier transform infrared spectroscopy (FTIR) data. Spectroscopic evidence for Cu agglomeration on SiO_2 is presented involving the intensity variations of the SiO_2 LO/TO phonon modes after chemical reaction with the Cu precursor and after the H_2 precursor cycle. These intensity variations are observed over the first 20 ALD cycles at 185°C.

INTRODUCTION

Atomic layer deposition (ALD) has recently received great interest with the need for new materials and thin-film deposition techniques for device scaling in integrated circuits (IC). The self-limiting growth mechanism of ALD is ideal for producing very thin, conformal films even on surfaces with high aspect ratios with control of the thickness and composition at the atomic level. As the industry heads toward the 22 nm node, the interest in metal ALD is growing, especially for copper interconnects. Cu is replacing aluminum as an interconnect material in ICs due to its lower resistivity ($1.72\times10^{-6}\Omega\cdot cm$ vs. $2.82\times10^{-6}\Omega\cdot cm$) and higher melting point (1083 °C vs. 659°C).[1] For this application, a highly conformal and continuous copper seed layer is required before subsequent electrochemical deposition of copper film with high growth rate.[2]

ALD of non-noble metals, however, has had limited success [3, 4]. One of the reasons for these difficulties is related to a lack of suitable precursors which satisfy stringent requirements for ALD. [5] In addition, the lack of understanding of growth mechanisms of metal ALD hinders the development of suitable precursors.

ALD of Cu has been considered previously. Several copper (I) and copper (II) precursors were commonly used in the past, such as $CuCl$[4, 6], Cu(II)-2,2,6,6 -tetramethyl-3,5-heptandionate $[Cu(thd)_2]$[7, 8], Cu(II)-1,1,1,5,5,5-hexafluoro-2,4 -pentanedionate $[Cu(hfac)_2]$ [9, 10], and Cu(II) acetylacetonate $[Cu(acac)_2]$ [11, 12], but all suffered from undesirable properties as ALD precursors.[13] For example, CuCl has very low vapor pressure, and the Cu film deposited by Cu(hfac)$_2$ contains fluorine which reduces the Cu adhesion on the substrate. They all have very low reactivity with low growth rate, so that either high temperature (> 200°C), which is not well suited for smooth Cu growth due to Cu agglomeration and diffusion [13-15], or plasma is needed to enhance the reactivity.

Recently, Gordon et al.[16] reported ALD growth of metallic copper with high conformality and high conductivity using an amidinate precursor (copper (I) di-sec-butylacetamidinate, [Cu(^{s}Bu-amd)]$_2$) and molecular hydrogen at relatively low temperature (~185°C). One of the advantages of this precursor is its high reactivity with molecular hydrogen at reasonable temperatures. In order to understand the ALD Cu growth mechanisms with this amidinate precursor, we used in-situ Fourier transform infrared spectroscopy (FTIR) to investigate the chemical bonding evolution after each ALD cycle.

EXPERIMENT

Double side polished, float-zone grown, and lightly doped (~10 Ω·cm) Si(100) with thin thermal oxide (6-10 nm thick SiO$_2$) is used. The sample is first rinsed by acetone, methanol, and deionized water (DI water, 18.2 MΩ·cm), then standard RCA [17, 18] cleaning is performed followed by thorough rinsing with deionized water and blow-drying with N$_2$. Then the sample is immediately loaded in the ultra pure N$_2$ (oxygen impurity < 10^{-3} ppm) purged ALD chamber.

All experiments are done in a home-made ALD system integrated with a Thermo Nicolet 670 interferometer and a MCT/B detector for *in-situ* FTIR measurements.[19] A single-pass transmission geometry is used with an incidence angle close to the Brewster angle to maximize transmission, minimize interference, and increase sensitivity to absorptions below ~1500 cm^{-1}. Distinction of perpendicular from parallel modes to the surface is made by additional normal incidence measurements. Mid-infrared range of 400 - 4000 cm^{-1} is scanned with the resolution of 4 cm^{-1}.

[Cu(^{s}Bu-amd)]$_2$ is kept at 95-100°C, and purified N$_2$ is used as the carrier gas to deliver the copper precursor. Cu film is deposited by introducing [Cu(^{s}Bu-amd)]$_2$ (exposure ~ 10^7 L) and H$_2$ (exposure ~ 10^{11} L) (ultra high purity, 99.999%, purified by Aeronex Gate Keeper gas purifiers) alternatively into the ALD chamber. During deposition, the substrate temperature is kept at 185°C for optimum Cu deposition. After each precursor dosing, the ALD chamber is N$_2$-purged thoroughly with pumping.

RESULTS AND DISCUSSION

Figure 1 shows differential IR absorbance spectra of the first cycle of Cu ALD on SiO$_2$ at 185°C, where the maximum Cu growth rate is observed.[13] Upon the first half cycle when the Cu precursor [Cu(^{s}Bu-amd)]$_2$ (black) is introduced, the intensity of the transverse and longitudinal optical modes (TO/LO) of SiO$_2$ at 1075/1247 cm^{-1} decreases. Their intensity loss indicates that the SiO$_2$ matrix is chemically disrupted due to the reaction of the Cu precursor with surface OH groups. The absorption band at 1010 cm^{-1} corresponds to the formation of Si-O-Cu bonds due to the adsorbed Cu precursor on SiO$_2$. The reaction scheme is shown as the inset in the right. Upon the H$_2$ exposure (red), there is a partial recovery of SiO$_2$ phonon modes (1075/1247 cm^{-1}). The recovery of these modes suggests that after the H$_2$ dosing, the Cu precursor on the surface is reduced to pure copper atoms which then migrate and self-agglomerate to form copper particles

[13, 21, 22]. As they agglomerate, the chemical bonds established during the first half cycle between Cu and the underlying SiO_2 matrix are broken and rehydroxylated, partially restoring the SiO_2 phonon modes and hydroxyl sites on the surface.

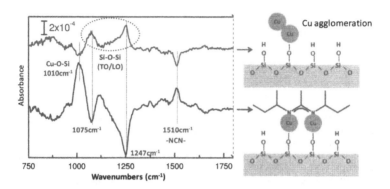

Figure 1. Differential IR absorbance spectra of the first cycle of Cu ALD on SiO_2 measured after the Cu precursor (black) and H_2 (red) dosing. Each spectrum is referenced to the spectrum of the previous treatment and the bottom is referenced to initial oxide surface. The incident angle of IR beam is $74°$. The inset in the right shows a schematic of surface reactions corresponding to observations in each spectrum.

In addition, the removal of the ligands by H_2 reduction opens more reaction sites which were initially blocked by the intact ligands. In addition to the repetitive change of SiO_2 phonon modes, the surface Si-O-Cu bonds at 1010 cm^{-1} follows the similar pattern of gain and recovery after Cu precursor and H_2, respectively. This process is again schematically illustrated in the inset of Figure 1. The repetitive loss and gain of SiO_2 modes are observed for the 20th cycle, implying that the surface is not saturated and Cu atoms still agglomerate after 20 ALD cycles. This observation is consistent with the high percolation thickness for Cu on glass reported previously.[13]

In Figure 1, the mode at 1510 cm^{-1} is assigned to $v(N-C-N)$ of the intact ligand attached to Cu atoms on the surface. Along with CH_x stretching modes at 2800-3000 cm^{-1} (not shown), their intensity increases after each copper precursor and decreases after each H_2, confirming the ALD process through ligand exchange.

Cu agglomeration (i.e. formation of Cu structures with substantially thicker diameter than the average Cu thickness) is also observed in the shape of Rutherford backscattering spectroscopy (RBS) of Cu in Figure 2. The areal density of Cu atoms is measured by ex situ RBS with 2 MeV He^+ ions. The detector is placed at $160°$ backscattering angle.

43

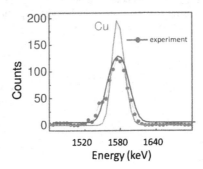

Figure 2. RBS spectrum (red dots) after 10 ALD cycles and simulations assuming uniform Cu coverage (gray) and non-uniform Cu coverage (blue). The detector is placed at a grazing angle of 160° backscattering angle.

The FWHM of the simulation assuming uniform coverage of Cu atoms on the surface (gray) is narrower than that of the experimental RBS data (red dots). The blue line shows the simulated result after taking into account agglomerated Cu atoms of an island structure on the surface. The simulation of non-uniform Cu coverage is reasonably well fitted to the RBS data, confirming the agglomerated Cu on SiO_2.

CONCLUSIONS

The surface reaction and the growth mechanism of Cu ALD with a liquid amidinate Cu precursor ($[Cu(sBu-amd)]_2$) and molecular hydrogen on oxidized Si surfaces (6-10 nm thick SiO_2) have been investigated by using in situ FTIR spectroscopy and ex situ RBS. The agglomeration of Cu atoms on SiO_2 was shown to have a significant effect on the evolution of SiO_2 phonon modes. The hydrogen reduction of the surface ligand is observed through the modes attributed to CH_x and -NCN- bonds, confirming ALD process through ligand exchange. The Cu agglomeration on the oxide is also confirmed by a broader RBS spectrum of Cu.

ACKNOWLEDGMENTS

This work was supported by the National Science Foundation (CHE-0415652).

REFERENCES

1.   The International Technology Roadmap for Semiconductors, Semiconductor Industry Association. http://pulic.itrs.net. 2005.
2.   Jae Jeong, K.; Soo-Kil, K.; Chang Hwa, L.; Yong Shik, K., Investigation of various copper seed layers for copper electrodeposition applicable to ultralarge-scale integration interconnection. Journal of Vacuum Science & Technology B 2003, 21, (1), 33-38.

3.   Lim, B. S.; Rahtu, A.; Gordon, R. G., Atomic layer deposition of transition metals. *Nat Mater* **2003**, 2, (11), 749-754.

4.   Marika, J.; Mikko, R.; Markku, L., Deposition of copper films by an alternate supply of CuCl and Zn. *Journal of Vacuum Science & Technology A* **1997**, 15, (4), 2330-2333.

5.   Musgrave, C.; Gordon, R. G., Precursors for Atomic Layer Deposition of High-K Dielectrics. *Future Fab International* **2005**, 18, 126-128.

6.   Martensson, P.; Carlsson, J.-O., Atomic Layer Epitaxy of Copper on Tantalum. *Chemical Vapor Deposition* **1997**, 3, (1), 45-50.

7.   Per, M.; Jan-Otto, C., Atomic Layer Epitaxy of Copper. *Journal of the Electrochemical Society* **1998**, 145, (8), 2926-2931.

8.   Christopher, J.; Lanford, W. A.; Christopher, J. W.; Singh, J. P.; Pei, I. W.; Jay, J. S.; Toh-Ming, L., Inductively Coupled Hydrogen Plasma-Assisted Cu ALD on Metallic and Dielectric Surfaces. *Journal of the Electrochemical Society* **2005**, 152, (2), C60-C64.

9.   Raj, S.; Balu, P., Atomic Layer Deposition of Copper Seed Layers. *Electrochemical and Solid-State Letters* **2000**, 3, (10), 479-480.

10.   Huo, J.; Solanki, R.; McAndrew, J., Characteristics of copper films produced via atomic layer deposition. *Journal of Materials Research* **2002**, 17, 2394-2398.

11.   Utriainen, M.; Kröger-Laukkanen, M.; Johansson, L.-S.; Niinist, L., Studies of metallic thin film growth in an atomic layer epitaxy reactor using M(acac)2 (M=Ni, Cu, Pt) precursors. *Applied Surface Science* **2000**, 157, (3), 151-158.

12.   Antti, N.; Antti, R.; Timo, S.; Kai, A.; Mikko, R.; Markku, L., Radical-Enhanced Atomic Layer Deposition of Metallic Copper Thin Films. *Journal of the Electrochemical Society* **2005**, 152, (1), G25-G28.

13.   Li, Z.; Rahtu, A.; Gordon, R. G., Atomic Layer Deposition of Ultrathin Copper Metal Films from a Liquid Copper(I) Amidinate Precursor. *Journal of Electrochemical Society* **2006**, 153, C787-C794.

14.   Benouattas, N.; Mosser, A.; Raiser, D.; Faerber, J.; Bouabellou, A., Behaviour of copper atoms in annealed Cu/SiOx/Si systems. *Applied Surface Science* **2000**, 153, (2-3), 79-84.

15.   McBrayer, J. D.; Swanson, R. M.; Sigmon, T. W., Diffusion of Metals in Silicon Dioxide. *Journal of the Electrochemical Society* **1986**, 133, (6), 1242-1246.

16.   Li, Z.; Barry, S. T.; Gordon, R. G., Synthesis and Characterization of Cu(I) Amidinates as Precursors for Atomic Layer Deposition (ALD) of Copper Metal. *Inorganic Chemistry* **2005**, 44, 1728-1735.

17.   Higashi, G. S.; Chabal, Y. J., Silicon surface chemical composition and morphology, Chapter in Handbook of Silicon Wafer Cleaning Technology: Science, Technology, and Applications *Werner Kern ed, Noyes Pub.,* **1993**.

18.   Weldon, M. K.; Marsico, V. E.; Chabal, Y. J.; Hamann, D. R.; Christman, S. B.; E. E. Chaban, S. S., Infrared Spectroscopy as a Probe of Fundamental Processes in Microelectronics: Silicon Wafer Cleaning and Bonding, . *Surface Science* **1996**, 368, 163.

19.   Kwon, J.; Dai, M.; Langereis, E.; Halls, M. D.; Chabal, Y. J.; Gordon, R. G., In-situ Infrared Characterization during Atomic Layer Deposition of Lanthanum Oxide. *Journal of Physical Chemistry C* **2009**, 113, (2), 654-660.

20.   Conley, R. T., Infrared Spectroscopy. *Allyn and Bacon, Inc.* **1972**.

21.   Yang, C.-Y.; Jeng, J. S.; Chen, J. S., Grain growth, agglomeration and interfacial reaction of copper interconnects. *Thin Solid Films* **2002**, 420-421, 398-402.

22.   Ching-Yu, Y.; Chen, J. S., Investigation of Copper Agglomeration at Elevated Temperatures. *Journal of the Electrochemical Society* **2003,** 150, (12), G826-G830.

Mater. Res. Soc. Symp. Proc. Vol. 1155 © 2009 Materials Research Society   1155-C04-04

Novel Hf and La Formamidinate Precursors for ALD Application

Huazhi Li and Deo V. Shenai[*]

Metalorganic Technologies, Dow Electronic Materials,
60 Willow Street, North Andover, MA 01845, U.S.A.

INTRODUCTION

Higher dielectric constant materials are required in order to continue scaling down gate dimensions and to eliminate the high current leakage experienced by ultrathin (<1.5 nm) films in metal oxide semiconductor field-effect transistor (MOSFET). Hafnium oxide has been used in 32 nm node, because of its exceptional properties such as high dielectric constant, excellent thermal stability and compatibility with Si processing[1]. La capped HfO_2 and HfSiO films have been reported for improving threshold voltage control[2]. The La incorporation in these HfO_2 and $HfSiO_x$ films has evidently improved the electrical performance [3] of these films. In the gate oxide application, atomic layer deposition (ALD) is particularly suitable due to its excellent conformity and precise thickness control. ALD also offers the best electrical characteristics in the device. In order to achieve ideal ALD performance, it is critical to use appropriate enabling precursors. Tetrakisethylmethylamidohafnium (TEMAHf) has been considered as one of the most promising precursors due to its good physical properties. However, this precursor has relatively low thermal stability, which has often become a drawback to ALD process for HfO_2. To circumvent the decomposition issues of TEMAHf during ALD deposition, often low process temperatures have been used that lead to less dense films and high level of carbon contamination. In order to overcome these challenges, the novel hafnium formamidinate (Hf-FAMD) precursor is developed as an alternate Hf source at Dow Electronic Materials. Following our success with formamidinate platform in previously reported novel lanthanum formamidinate (La-FAMD) source[4], this new precursor also exhibits high thermal stability and high reactivity towards water and ozone. In this work, we report that Hf-FAMD exhibits acceptable vapor pressure (> 0.1 Torr at 100 °C), and higher thermal stability for ALD applications. We also present the crystal structure of La-FAMD, and physical properties of novel Hf-FAMD relevant to ALD.

EXPERIMENTAL

Hf-FAMD was synthesized according to the synthetic strategy reported for metal formamidinates in our earlier work[4]. All product characterizations and measurements were performed under anaerobic conditions, using a blanket of inert N_2 gas. The details of characterizations are elaborated below:

The high purity of precursor was confirmed by using Proton Nuclear Magnetic Resonance (^1H NMR) spectroscopy and Inductively Coupled Plasma – Mass Spectrometry (ICP-MS) analytical techniques. The analysis of Hf-FAMD product, after purification, confirmed the absence of organic impurities (such as trace solvents or reagents) below 50 ppb by FT-NMR, and while the metallic impurities to be below 10 ppb level by ICP-MS. Thermogravimetric analysis (TGA) for Hf-FAMD was carried out with TA Instruments TA Q50 TGA Analyzer kept inside a nitrogen glove box. The sample N_2 flow rate was 60 sccm. Samples (10-20 mg) were loaded in open platinum crucibles following standard procedures. The measurements were carried out with a temperature ramp rate of 10 K/min. The melting points were determined using sealed capillaries and Mel-Temp II melting point apparatus. The vapor pressure – temperature curve, for Hf-FAMD saturated vapors, was established using static vapor pressure measurement technique reported earlier in the literature[5]. The thermal stability studies were carried out using Accelerating Rate Calorimetry (ARC), and typically 2 grams of sample. The onset of decomposition was determined by the self-heating mechanism during the ARC testing. The crystal structure was elucidated using Single Crystal X-Ray Crystallography, using the pure colorless prisms of La-FAMD crystallized from its pentane solution. A crystal of 0.207 mm x 0.237 mm x 0.296 mm in size was mounted on a nylon loop with Paratone-N oil, and transferred to a Bruker SMART APEX II[R] diffractometer equipped with an Oxford Cryosystems 700 Series Cryostream Cooler and Mo Kα radiation (λ = 0.71073 Å). A total of 3066 frames were collected at 193 (2) K to θ_{max} = 30.00°, with an oscillation range of 0.5°/frame and an exposure time of 20 s/frame. The specialized softwares employed for crystal structure elucidation include APEX[6], SAINT[7], SADABS[8], and SHELXTL[9] at Harvard University, MA.

RESULTS AND DISCUSSIONS

Previously we had disclosed the novel La-FAMD precursor and its application in the deposition of LaAlO₃ thin films[4] by ALD. This La precursor was found to exhibit exceptionally high thermal stability, enabling a wider process temperature window for ALD. Also the carbon impurity incorporation was found to be negligible while using this precursor [4]. Recently, high quality LaAlO₃[4, 10] and LaLuO₃ films[11] have been grown using this precursor. As of today, it remains the most volatile La source, with vapor pressure of 82 mTorr at 120 °C. These interesting properties of La-FAMD have prompted us to study its crystal structure in more detail.

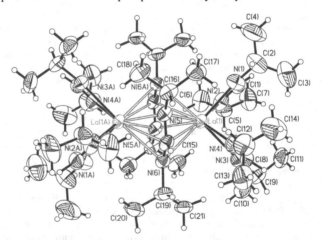

Figure 1. The molecular structure of La-FAMD with hydrogens, showing the disorder present in the isopropyl moieties. The non-hydrogen atoms are depicted with 50% probability ellipsoids.

Table 1. Crystallographic data for La-FAMD

formula	$C_{42} H_{90} La_2 N_{12}$	β	126.137(1)°
formula weight	1041.08	V	2654.77(5) Å^3
crystal system	Monoclinic	d_{cacl}	1.302 g/cm^3
space group	P2$_1$/c (No. 14)	μ	1.625 mm^{-1}
Z	2	θ range	1.89 to 30.00°
a	11.1019(1) Å	Refinement method	Full-matrix least-squares on F^2
b	17.2787(2) Å	GOF (F^2)	1.290
c	17.1363(2) Å	R_1^b (wR_2^c),%	4.61(8.63)

Table 2. Selected bond lengths (Å) and angles (deg) for La-FAMD

La(1)-N(1)	2.485(3)	La(1)-N(6A)	2.857(2)
La(1)-N(2)	2.613(3)	La(1)-La(1A)	3.8943(3)
La(1)-N(3)	2.530(3)	N(1)-La(1)-N(2)	54.02(12)
La(1)-N(4)	2.583(3)	N(3)-La(1)-N(4)	53.45(11)
La(1)-N(5)	2.679(2)	N(5)-La(1)-N(6)	47.89(8)
La(1)-N(6)	2.869(2)	N(5A)-La(1)-N(6A)	48.18(7)
La(1)-N(5A)	2.658(2)		

The crystal structure of La-FAMD is shown in Figure 1. The asymmetric unit contains half of the molecule with the center located in the middle point of La(1)-La(1A) axis. Specific crystallographic parameters are listed in Table 1 and some selected bond lengths and angles are listed in Table 2. The average La-N$_{non-bridging}$ distances are found to be 2.55 Å, which are close to what was reported before for lanthanum tris(N,N'-diisopropyl-2-tert-butylamidinate) (2.53 Å)[12] and for lanthanum tris(N,N'-ditertbutylacetamidinate) (2.54 Å)[13]. The chelating angles between La-N$_{non-bridging}$ bonds are found to average at 53.7°. These values are slightly larger than (a) 51.8° reported for lanthanum tris(N,N'-diisopropyl-2-tert-butylamidinate) and (b) 52.2° reported for lanthanum tris(N,N'-ditertbutylacetamidinate). This can be attributed to the fact that formamidinate ligand is sterically less bulky than analogous acetamidinate. Interestingly, average La-N$_{bridging}$ bond distances (2.77 Å) are significantly longer than those of La-N$_{non-bridging}$ (2.55 Å). Also worthwhile to note that the chelating angles between La-N$_{bridging}$ bonds (48.0°) are smaller than those of La-N$_{non-bridging}$ bonds (53.7°). Even though the crystal study shows the material is a dimer in solid state, it is also the most volatile La source developed so far. We believe that this exceptional high vapor pressure is the result of possible equilibrium exchange between the dimer and monomer upon heating. This hypothesis is substantiated by the FT-NMR studies[4], which indicate that the dissociation of La-FAMD dimer into monomer occurs readily in solution leading to equilibrium between monomer and dimer at room temperature. At higher temperatures exceeding 60 °C, the ratio of monomer over dimer becomes extremely high and approaches close to infinity with increased temperatures beyond 100 °C. Conversely, this phenomenon can be explained by the crystallography results for La-FAMD. To illustrate, since the two La centers are too far away (3.89 Å) and La-N$_{bridging}$ is significantly longer than La-N$_{non-bridging}$, the dissociation is facilitated readily at either sublimation temperatures or with dissolution.

As a recent development, La-FAMD has shown very promising properties as a dopant candidate for high dielectric and metal gate application, involving Hf-based materials. This new discovery has triggered the researchers to develop a new Hf precursor that offers the appropriate process window matching with La-FAMD based processes. Since the La-FAMD has demonstrated greater thermal stability, we have strategically extended the same formamidinate platform and chemistry to Hf. The new Hf-FAMD precursor was synthesized and evaluated at Dow Electronic Materials in the same manner as La-FAMD. Using the proprietary production process, this new precursor is now made available to meet the customer needs.

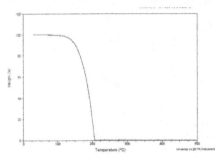

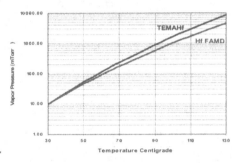

Figure 2. TGA of Hf-FAMD   **Figure 3.** Vapor pressures of Hf Sources

The melting point of Hf-FAMD was found to be 141 °C, while the Thermogravimetric Analysis (TGA) showed a clean sublimation with negligible residue (less than 1%) as shown in Figure 2. The temperature for 50% mass loss (often denoted as $T_{1/2}$) was found to be around 180 °C. These results confirm the greater thermal stability of Hf-FAMD vis-à-vis conventional amido source, TEMA-Hf, and also validate the evaporation rate to be acceptable for ALD process.

The temperature dependence of Hf-FAMD saturated vapor pressure was studied using static vapor pressure measurement method [5] developed specially for air-sensitive metalorganics. A side-by-side comparison of vapor pressures was conducted using TEMAHf. The accuracy of the Baratron™ pressure transducer system is at ± 0.001 Torr for the pressure measurements under consideration. After accurate determinations of vapor pressures at various temperatures, followed by their reproducibility confirmations, the vapor pressure equations were established as **log P (Torr) = 8.72 – 3248/T**, and **log P (Torr) = 9.86 –3599/T** for Hf-FAMD and TEMAHf respectively. Figure 3 shows the comparison of vapor pressure curves for both hafnium sources, and confirms that Hf-FAMD has comparable vapor pressure to TEMAHf over a wide range of temperatures, which makes Hf-FAMD suitable and safer (being a solid) source for ALD use.

Thermal stability is another important criteria to consider while designing new ALD precursors. We have verified the thermal stability of Hf-FAMD by two independent and corroborating techniques: one is to subject the precursor and its solution to constant high temperature (200 °C) over a prolonged period, and then to look for the sign of decomposition by sequentially studying its ^{1}H NMR spectrum; another is to use ARC to determine the thermal stability. A sealed NMR tube, containing Hf-FAMD dissolved in d^6-C_6H_6, was heated to 200 °C for a prolonged period, and the ^{1}H NMR studies were carried out at set intervals during the heating cycle. No sign of decomposition or extra peaks were found over 1200 hours, see e.g., percentage precursor remaining over time for both sources are shown in Figure 4. These results indicate excellent thermal stability. On the contrary, total decomposition of TEMAHf was found to occur within 10 hours under identical conditions. These results clearly confirm the thermal stability benefit of Hf-FAMD over conventional source TEMAHf. The thermal stability studies using ARC was another technique employed in this study. This technique can determine the temperature at which a self-heating or runaway reaction can start. Typically the sample (around 2 g) is placed in a spherical sample cell, which is then placed in the ARC calorimeter. Figure 5 highlights the differences observed in ARC studies of Hf-FAMD and TEMAHf. Hf-FAMD was thermally stable up to the experimental limit (300 °C) while TEMAHf started to decompose at 200 °C. The thermal advantage of Hf-FAMD over TEMAHf suggests that Hf-FAMD precursor can be used at higher temperature without decomposition, so that lower C impurity incorporation can be expected in the final films. Also it can be easily integrated into La process based on La-FAMD since they both have the similar and matching thermal properties.

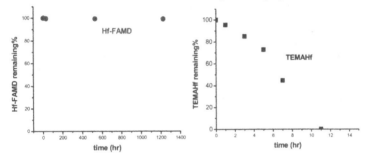

Figure 4. Thermal decomposition curves of Hf-FAMD (left) and TEMAHf (right).

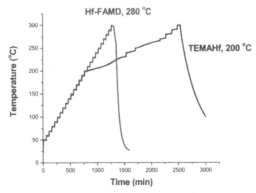

Figure 5. ARC study on Hf-FAMD and TEMAHf

Preliminary ALD studies using solid Hf-FAMD and water or ozone have resulted in acceptable HfO_2 films. The refractive index of these initial films is around 2.1 and the growth rate is found to be around 0.8 Å/cycle, acceptable for ALD. Further studies and commercial evaluations are currently in progress.

CONCLUSIONS

The crystal structure of La-FAMD was resolved to confirm that La-FAMD is indeed a dimer, which dissociates into monomeric form in gas phase and solution. A novel Hf precursor using formamidinate platform, viz Hf-FAMD was developed and produced in large quantities. Hf-FAMD exhibits significantly greater thermal stability than commercially available Hf precursors such as TEMAHf. It also showed acceptable volatility (60 mTorr at 100°C), which is comparable with that of TEMAHf. The preliminary ALD studies indicated that both La-FAMD and Hf-FAMD offer similar temperature process window, which should essentially enable their integration easier. Preliminary deposition of La and Hf oxide films using new formamidinate precursors confirmed the high reactivity of these sources towards water and ozone alike.

ACKNOWLEDGMENTS

We thank Professor Roy G. Gordon, Harvard University, for useful discussions and productive collaboration. The authors would also like to thank Dr. Douglas Ho from Harvard University, and Drs. Daewon Hong, Jean-Sebastien Lehn and Joseph Magee from Dow Advanced Materials, for their valuable technical contributions.

REFERENCES

1. R. Chau; S. Datta; M. Doczy; B. Doyle; J. Kavalieros; M. Metz, IEEE Elect. Dev. Lett., **2004**, 25(6), 408-410.
2. S. Kamiyama; T. Miura; E. Kurosawa; M. Kitajima; M. Ootuka; T. Aoyama; Y. Nara, *Symp. VLSI Tech.*, **2007**, 539-542.
3. H. N. Alshareef, H. R. Harris, H. C. Wen, C. S. Park, C. Huffman, K. Choi, H. F. Luan, P. Majhi, B. H. Lee, R. Jammy, D. J. Lichtenwalner, J. S. Jur, A. I. Kingon, *Symp. VLSI Tech.*, **2006**, 10.
4. H. Li, D. Shenai, R. Pugh, J. Kim, *Mater. Res. Soc. Symp. Proc.* **2008**, 1036E, 1036-M04-18.
5. D. V. Shenai, M. L. Timmons, R. L. DiCarlo, G. K. Lemnah, R. S. Stennick, *J. Crystal Growth*, **2003**, 248, 91.
6. Bruker AXS (2006a). *APEX2 v2.1-0*. Bruker AXS Inc., Madison, Wisconsin, USA.
7. Bruker AXS (2006b). *SAINT V7.34A*. Bruker AXS Inc., Madison, Wisconsin, USA.
8. Bruker AXS (2004). *SADABS*. Bruker AXS Inc., Madison, Wisconsin, USA.
9. Bruker AXS (2001). *SHELXTL v6.12*. Bruker AXS Inc., Madison, Wisconsin, USA.
10. Y. Liu, H. Kim, J. J. Wang, H. Li, R. G. Gordon, *ECS Transactions* **2008**, (16)5, 471-478.
11. H. Wang, J. J. Wang, R. G. Gordon, J. S. Lehn, H. Li, D. Hong, D. Shenai, *Electrochem. Solid-State Letts.*, **2009**, 12(4), G13-G15.
12. B. S. Lim, A. Rahtu, J.-S. Park, R. G. Gordon, *Inorg. Chem.*, **2003**, 42, 7951-7958.
13. J. Päiväsaari, C. L. Dezelah IV, D. Back, H. M. El-Kaderi, M. J. Heeg, M. Putkonen, L. Niinistö, and C. H. Winter, *J. Mater. Chem.*, **2005**, *15*, 4224-4233.

Mater. Res. Soc. Symp. Proc. Vol. 1155 © 2009 Materials Research Society 1155-C07-05

Impact of Post Deposition Annealing on Characteristics of $Hf_xZr_{1-x}O_2$

D.H. Triyoso, R.I. Hegde, R. Gregory, G.S. Spencer and W. Taylor Jr.
Technology Solutions Organization, Freescale Semiconductor Inc.
3501 Ed Bluestein Blvd, Austin, TX 78721, U.S.A.

ABSTRACT

In this paper the impact of post deposition annealing in various ambient on electrical properties of hafnium zirconate ($Hf_xZr_{1-x}O_2$) high-k dielectrics is reported. ALD $Hf_xZr_{1-x}O_2$ films are annealed in a nitrogen and/or oxygen ambient at 500°C to 1000°C. Devices annealed at 500°C in N_2 has lower equivalent oxide thickness (EOT) of 10Å without significant increase in gate leakage (J_g), threshold voltage (V_t) and only a slight decrease in transconductance (G_m) values compared to 500°C O_2 annealed devices. Furthermore, the impact of annealing $Hf_xZr_{1-x}O_2$ films in a reducing ambient (NH_3) is studied. Optimized NH_3 anneal on $Hf_xZr_{1-x}O_2$ results in lower CET, improved PBTI, low sub-threshold swing values, comparable high-field G_m with only a minor degradation in peak G_m compared to control $Hf_xZr_{1-x}O_2$. Finally, the impact of laser annealing vs. RTP annealed $Hf_xZr_{1-x}O_2$ films are reported. Laser annealing helped further stabilize tetragonal phase of $Hf_xZr_{1-x}O_2$ without inducing void formation. Good devices with low leakage, low EOT and high mobility are obtained for laser annealed $Hf_xZr_{1-x}O_2$.

INTRODUCTION

Hafnium based high-k dielectrics have recently been implemented by a few companies for 45nm technology node. Most companies, however, have pushed back high-k implementation until 32 nm technology node. For successful implementation at 32 nm and beyond, extending scalability of hafnium based high-k dielectrics is needed. We have recently reported stabilization of tetragonal phase of HfO_2 by zirconium addition resulting in improved scalability of hafnium-based dielectrics [1-3]. This paper reports the impact of post deposition annealing in various ambient on electrical properties of hafnium zirconate high-k dielectrics.

EXPERIMENT

ALD $Hf_xZr_{1-x}O_2$ films (x ~0.4) of varying thickness were annealed in a nitrogen and/or oxygen ambient at temperatures ranging from 500°C to 1000°C. All films were formed via atomic layer deposition (ALD) using hafnium tetrachloride, zirconium tetrachloride, and deuterated water at a deposition temperature of 300°C. Film thickness is controlled by the number of cycles deposited. The dielectric layer was grown on a chemical oxide starting surface. The chemical oxide was formed by cleaning Si wafers in a solution of de-ionized water, hydrogen peroxide and hydrochloric acid. For electrical characterization, the ~ 30Å high-k dielectric deposition is followed with a TaC_y metal gate electrode and capped with poly-Si. The high-k dielectric films were deposited on 'cake oxide' wafers (wafers with 0, 50 and 100Å SiO_2 in concentric rings) [3]. The NMOS long channel transistors were fabricated using conventional CMOS integration with sidewall liners and spacers, implants to the Si-cap, implants to source/drain, 1000°C activation anneal, cobalt-salicide contacts, and forming gas anneal. Equivalent oxide thickness (EOT), threshold voltage (V_t) and flatband voltage (V_{fb}) were extracted using the Hauser program. 85 x 80 μm^2 capacitors and 10 x10 μm^2 transistors were then probed electrically.

RESULTS AND DISCUSSION

Impact of O_2 and N_2 Anneal

Figure 1a shows capacitance-voltage (C-V) plot for $Hf_xZr_{1-x}O_2$ NMOS devices with various post deposition annealing conditions. All post deposition annealing conditions resulted in well-behaved C-V curves. Each set of curves are representative of over 40 measurements. The devices with the 500°C 60s O_2 anneal has the highest EOT of ~10.9Å. When annealing ambient changed from O_2 to N_2, a ~ 0.9Å decrease in EOT is observed. We believe a reduction in interfacial layer occurs when nitrogen instead of oxygen is used, thus resulting in a decrease in overall EOT. A decrease in EOT (or an increase in dielectric constant) as a result of annealing high-k dielectrics in nitrogen ambient has been reported by others [4-5]. This EOT decrease has been shown due to a decrease in interfacial layer between high-k and Si substrate [4]. Keeping the annealing ambient the same (N_2), we look at the impact of annealing temperature on EOT. As the temperature is increased from 500°C to 700°C, a reduction in EOT is observed from ~ 10.0Å to 9.4Å. Further increase in annealing temperature to 800°C does not result in additional EOT decrease. The films annealed at 1000°C did not yield, most likely due to void formation. The AFM analysis performed on $Hf_xZr_{1-x}O_2$ films (deposited on un-patterned wafers) annealed at 1000°C in N_2 for 5s showed void formation [6]. Void formation on ALD hafnium-based high-k film after a high temperature anneal has also been reported by other groups [7]. A dual anneal scheme of 700°C 60s N_2 followed with 500°C 60s O_2 was attempted and resulted in lower EOT than O_2 only anneal (~10.5Å vs. 10.9Å) but higher EOT than N_2 only anneal (~ 9.4Å). All of these results indicate that 700°C 60s N_2 is an optimum annealing condition with regards to EOT reduction. However, there are additional factors that need to be considered when optimizing the impact of post deposition annealing on electrical characteristics such as J_g, V_t values and G_m.

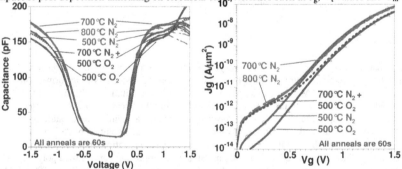

Figure 1. (a) Capacitance-voltage and (b) Leakage current density – voltage of 80x85μm^2 $Hf_xZr_{1-x}O_2$ devices with various post deposition annealing conditions.

Figure 1b shows J_g curves for $Hf_xZr_{1-x}O_2$ devices with various post deposition annealing conditions. As annealing ambient is changed from O_2 to N_2, a slight increase in leakage is observed, from 8.7×10^{-10} to 2.3×10^{-9} $A/\mu m^2$. No significant increase in leakage current density is observed as annealing temperature is increased to 800°C. The device annealed at 1000°C is very leaky due to void formation as evidenced by AFM results. The devices with dual anneal (700°C N_2/ 500°C O_2) have slightly lower leakage than devices with 700°C N_2 anneal for Vg>0.5V and the same leakage as devices with 700°C N_2 anneal for Vg<0.5V.

Figure 2 plots EOT vs. V_t for $Hf_xZr_{1-x}O_2$ devices with various post deposition annealing conditions. The lowest V_t's are obtained for film with 500°C anneal, regardless of annealing ambient (O_2 vs. N_2). Increasing the annealing temperature results in a small increase in V_t. Only 20-40 mV increase V_t is observed when annealing temperature is increased from 500°C to 700°C. The devices with dual annealing (N_2/O_2) have the highest V_t. Changes in V_t's may be due to change in interfacial layer between high-k and Si. The extracted work-function and fixed charge values obtained from V_{fb} vs. EOT data are shown in Table 1. The work-function is unchanged with annealing condition. The fixed charge values are comparable for all working devices, $\sim 1\text{-}2 \times 10^{11} cm^{-2}$.

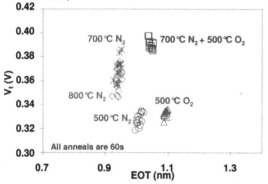

Figure 2. Plot of EOT vs. threshold voltage (V_t) of $10 \times 10 \mu m^2$ $Hf_xZr_{1-x}O_2$ devices with various post deposition annealing conditions.

Table 1. Summary of electrical properties of $Hf_xZr_{1-x}O_2$ devices with various post deposition annealing conditions. Gate leakage current density (Jg) is taken at 1V.

Post annealing	EOT (A)	J ($A/\mu m^2$)	Vt (V)	Wfn (eV)	Qf (cm^{-2})	Norm Gm (A/V/Å)
500 C 60s O_2 40T	10.9	8.7×10^{-10}	0.33	4.35	1.7×10^{11}	5.5×10^{-4}
500 C 60s N_2 10T	10.0	2.3×10^{-9}	0.33	4.35	2.3×10^{11}	5.0×10^{-4}
700 C 60s N_2 40T	9.4	3.5×10^{-9}	0.37	4.40	1.2×10^{11}	4.3×10^{-4}
800 C 60s N_2 40T	9.4	3.8×10^{-9}	0.36	4.39	1.3×10^{11}	4.2×10^{-4}
1000 C 5s N_2 740T	--	3.7×10^{-6}	--	4.50	5.8×10^{11}	--
700 C 60s N_2 40T / 500 C 60s O_2 40T	10.5	1.0×10^{-9}	0.39	4.41	1.9×10^{11}	4.7×10^{-4}

Figure 3 shows normalized Gm for $Hf_xZr_{1-x}O_2$ devices after various post deposition annealing. The devices annealed at 500°C with N_2 have lower G_m as those devices annealed at 500°C in O_2. As annealing temperature is increased from 500°C to 700°C, further decrease in peak G_m is observed. The Gm degradation can be mitigated by adding an oxygen anneal after nitrogen anneal as seen in the devices with dual anneal. Note that even though the oxygen anneal help improve Gm characteristics, the Gm is still lower than the device that receives oxygen anneal alone. We believe that reduction in interfacial layer thickness that results in EOT reduction after 700°C N_2 anneal negatively impact G_m characteristics.

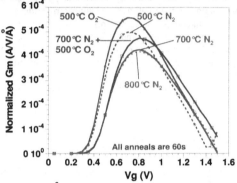

Figure 3. G_m of $10 \times 10 \mu m^2$ $Hf_xZr_{1-x}O_2$ transistors with various annealing conditions.

Impact of NH₃ Anneal

The impact of post-deposition annealing in a reducing ambient is studied by annealing ALD $Hf_xZr_{1-x}O_2$ films with varying thickness in NH_3 as shown in Figure 4. A significant reduction in CET to as low as 14Å (EOT~9Å) is observed due to NH_3 anneal as long as the $Hf_xZr_{1-x}O_2$ physical thickness is 15Å or thicker. When the dielectric thickness is reduced further, leakage current increases to an unacceptable level.

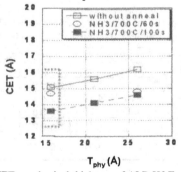

Figure 4. Plot of CET vs. physical thickness of ALD $Hf_xZr_{1-x}O_2$ films annealed in NH_3.

Optimized NH_3 anneal on $Hf_xZr_{1-x}O_2$ results in lower CET and good device characteristics. The detailed device characteristics have recently been reported [8]. Figure 5 shows NH_3 annealed devices with improved PBTI and comparable high-field Gm. There is only a minor degradation in peak Gm compared to control $Hf_xZr_{1-x}O_2$.

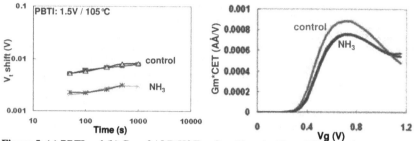

Figure 5. (a) PBTI and (b) Gm of ALD $Hf_xZr_{1-x}O_2$ with and without NH_3 anneal.

Impact of Laser Anneal

While post anneals are typically done at low temperature, the impact of post-annealing $Hf_xZr_{1-x}O_2$ films with laser vs. RTP was studied. Laser annealing was done at 1275-1325°C on the order of one millisecond in air and RTP annealing was done at 1000°C for 5s in nitrogen ambient. Laser annealing helped further stabilize tetragonal phase of $Hf_xZr_{1-x}O_2$ without inducing void formation as shown in Figure 6. Good devices with low leakage, low equivalent oxide thickness (EOT) and high mobility are obtained for laser annealed $Hf_xZr_{1-x}O_2$ as previously reported [9]. RTP annealed devices are leaky, most likely due to void formation induced by high temperature annealing.

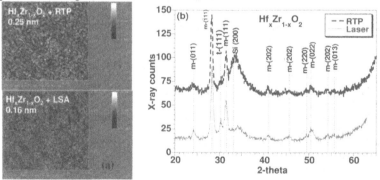

Figure 6. (a) AFM images and (b) XRD spectra of laser annealed vs. RTP annealed $Hf_xZr_{1-x}O_2$ films.

CONCLUSIONS

Impact of post deposition annealing in N_2 and/or O_2 for $Hf_xZr_{1-x}O_2$ indicate annealing in N_2 ambient results in lower EOT. A 1.5Å reduction is EOT is obtained by annealing at 700°C N_2 vs. 500°C O_2. The trade off of N_2 anneal is slightly higher gate leakage (which is a consequence of lower EOT) and lower G_m. Of all the N_2/O_2 annealing conditions studied, the best result is obtained by annealing at 500°C in N_2 ambient. At this temperature, lower EOT (10Å) is achieved without significant increase in J_g, V_t and only a slight decrease in G_m values compared to 500°C O_2 annealed devices. Further scaling of $Hf_xZr_{1-x}O_2$ can be achieved with NH_3 anneal. Optimized NH_3 anneal on $Hf_xZr_{1-x}O_2$ results in sub 10Å EOT, improved PBTI, low sub-threshold swing values, comparable high-field G_m with only a minor degradation in peak G_m compared to control $Hf_xZr_{1-x}O_2$. Furthermore, the impact of laser vs. RTP annealed $Hf_xZr_{1-x}O_2$ films are studied. Post-annealing with a laser helped further stabilize tetragonal phase of $Hf_xZr_{1-x}O_2$ without inducing the ruinous void formation observed with 1000°C 5s RTP anneals. Good devices with low leakage, low EOT and high mobility are obtained for laser annealed $Hf_xZr_{1-x}O_2$.

ACKNOWLEDGMENTS

The authors gratefully acknowledge Victor Wang for management support. The authors thank Jamie Schaeffer, Mark Raymond, Sri Samavedam, and David Gilmer for helpful discussion. The authors thank Darrell Roan, Liz Hebert, Jen-Yee Nguyen, Xiang-Dong Wang, Ross Noble, Eric Luckowski, and Chris Happ for technical assistance.

REFERENCES

1. D.H. Triyoso, R.I. Hegde, J.K. Schaeffer, D. Roan, P.J. Tobin, S.B. Samavedam, B.E. White, R. Gregory, and X.-D. Wang, Appl. Phys. Lett. 88, 222901 (2006).
2. R. I. Hegde, D.H. Triyoso. S. Samavedam and B.E. White Jr., J. Appl. Phys. 101 074113 (2007).
3. D.H. Triyoso, R.I. Hegde, J.K. Schaeffer, R. Gregory, X.-D. Wang, M. Canonico, D. Roan, E.A. Hebert, K. Kim, J. Jiang, R. Rai, V. Kaushik, and S.B. Samavedam, J. Vac. Sci. Technol. B. 25 845-52 (2007).
4. B.H. Lee, L. Kang, R. Nieh, W.-J. Qi, and J.C. Lee, Appl. Phys. Lett. 76, 1926 (2000).
5. S.W. Jeong and Y. Roh, J. Korean Phys. Soc. 50, 1865 (2007).
6. D.H. Triyoso, P.J. Tobin, B.E. White Jr., R. Gregory, and X.D. Wang, Appl. Phys. Lett. 89, 132903 (2006).
7. P.S. Lysaght, B. Foran, G. Bersuker, P.J. Chen, R.W. Murto, and H.R. Huff, Appl. Phys. Lett. 82 1266 (2003).
8. R.I Hegde and D.H. Triyoso, J. App. Phys. 104 p.094110-8 (2008).
9. D.H. Triyoso, G. Spencer, R.I. Hegde, R Gregory, X.-D Wang, Appl. Phys. Lett. 92 113501-1 (2008).

Mater. Res. Soc. Symp. Proc. Vol. 1155 © 2009 Materials Research Society   1155-C09-14

Adsorption and Reaction of Hf Precursor With Two Hydroxyls on Si (100) Surface: First Principles Study

Dae-Hyun Kim[1], Dae-Hee Kim[1], Hwa-Il Seo[2], Ki-Young Kim[1], and Yeong-Cheol Kim[1]
[1]Department of Materials Engineering, [2]School of Information Technology,
Korea University of Technology and Education, 1800 Chungjeollo, Byungchun-myun, Chonan, 330-708, South Korea

ABSTRACT

Density functional theory was used to investigate the adsorption and reaction of $HfCl_4$ with two hydroxyls on Si (001)-2×1 surface in atomic layer deposition (ALD) process. When H_2O molecules are adsorbed on Si (001) surface at room temperature, they are dissociated into hydrogens and hydroxyl groups. There are two dissociation pathways; inter-dimer dissociation and intra-dimer dissociation. The activation energies of these pathways can be converted to the reaction probabilities. The probability for inter-dimer dissociation is ca. 67 % and the probability for intra-dimer dissociation is ca. 33 %. We prepared a reasonable Si substrate which consisted of six inter-dimer dissociated H_2O molecules and two intra-dimer dissociated H_2O molecules. The $HfCl_4$ must react with two hydroxyls to be a bulk-like structure. There were five reaction pathways where $HfCl_4$ can react with two hydroxyls on; inter-dimer, intra-dimer, cross-dimer, inter-row, and cross-row. Inter-row, inter-dimer and intra-dimer were relatively stable among the five reaction pathways based on the energy difference. The electron densities between O and Hf in these three reactions were higher than the others and they had shorter Hf-O and O-O bond lengths than the other two reaction pathways. The electron density and Hf-O, O-O bond lengths had influence on the energy difference.

INTRODUCTION

The complementary metal oxide semiconductor (CMOS) device is the most important electronic device in microelectronic industry. Silicon dioxide (SiO_2) is used extensively as dielectric materials. However, SiO_2 is so thin that the tunneling leakage current becomes too high.[1] High-k materials have received much attention recently in microelectronics because high-k gate dielectrics can significantly suppress the tunneling leakage current in CMOS devices. Among various high-k materials, hafnium dioxide (HfO_2) is considered to be the most promising candidate due to its relatively high permittivity, good thermal stability, and compatibility with dual metal integration.[2] The great advantage of SiO_2 is that it can be grown by thermal oxidation. In contrast, high-k oxides must be deposited.[2] Atomic layer deposition (ALD) technique is a desirable process to form HfO_2 because it shows good conformality and uniformity over large areas.[3,4] Willis et al. studied a reaction of $HfCl_4$ with H_2O terminated Si (001)-2×1 using density functional theory calculation.[5] The model predicted a saturation coverage of $(2.05\pm0.05) \times 10^{14} Hf/cm^2$ and the main factor which limited the reactant coverage was found to be a preferred reaction at two hydroxyl sites. They demonstrated the surface made up of roughly equal numbers of $OHfCl_3$ and $(O)_2HfCl_2$. Tang et al. performed density functional theory calculation to investigate an interface state of $Si:HfO_2$.[6] They calculated several interface models of $Si:HfO_2$ and found the most favorable interface structure.

The HfCl$_4$ should react with two hydroxyls to form a bulk-like structure in ALD for HfO$_2$. In this study, we calculated the reaction of HfCl$_4$ with various two hydroxyls sites on Si (001) - 2×1 based on density functional theory.

CALCULATION

Density functional theory (DFT) calculations were performed using the Vienna *ab-initio* Simulation Package (VASP) code with the projector augmented wave (PAW) potentials and the generalized gradient approximation (GGA).[8-12] The residual minimization scheme direct inversion in the iterative subspace (RMM-DIIS) was used for calculating the ground state of electrons.[13,14] A cutoff energy was 450 eV for the plane wave expansion of the wave functions and Monk-horst pack k-point mesh 2×2×1 produced well converged results.

When H$_2$O molecules are adsorbed on Si (001) surface at room temperature, they are dissociated with hydrogens and hydroxyls. There are two dissociation pathways; inter-dimer dissociation and intra-dimer dissociation. The activation energies of these pathways could be converted to the reaction probabilities. It was approximately 2:1.[14] Figure 1 (a) shows H$_2$O terminated Si (001) surface which was formed by 6 inter-dimer dissociations and 2 intra-dimer dissociations.

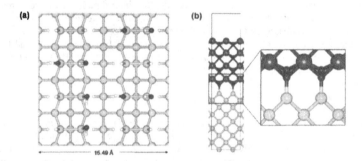

Figure 1. (a) H$_2$O terminated Si (001) surface which is formed by 6 inter-dimer dissociations and 2 intra-dimer dissociations and (b) an interface of tetragonal-HfO$_2$ and Si.

DISCUSSION

Figure 1 (b) shows an interface of tetragonal-HfO$_2$ and Si. This interface was the most stable among several interfaces of tetragonal-HfO$_2$ and Si.[15] Hf atoms located between two oxygen atoms which were on Si atoms. We concluded that HfCl$_4$ should react with two hydroxyls to be bulk-like structure. The reaction was calculated by the following equation.

$$HfCl_4 + 2\ OH\text{-}Si \rightarrow HfCl_2\text{-}O\text{-}Si + 2HCl$$

There were five reaction pathways of $HfCl_4$ for reacting with two hydroxyls; cross-dimer, cross-row, inter-dimer, intra-dimer, and inter-row. Figure 2 shows the configuration of each reaction pathways.

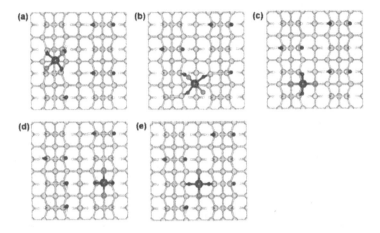

Figure 2. The reactions of $HfCl_4$ with two hydroxyls on Si (001) surface (a) cross-dimer, (b) cross-row, (c) inter-dimer, (d) intra-dimer, and inter-row.

Table 1. Bond lengths, bond angles, and energy differences of the each reaction.

	Bond length (Å)			Bond Angle (°)		ΔE (eV)
	Hf-O	O-O	Hf-Cl	∠ O-Hf-O	∠ Cl-Hf-Cl	
(a) Cross-dimer	1.97	3.35	2.38	115.78	97.13	0.79
(b) Cross-row	2.00	3.56	2.40	125.94	90.20	1.09
(c) Inter-dimer	1.93	3.06	2.34	105.23	114.35	0.30
(d) Intra-dimer	1.93	2.87	2.35	96.32	117.26	0.51
(e) Inter-row	1.93	3.19	2.34	111.55	108.65	0.31
$HfCl_4$ molecule			2.31		109.5	

Table 1 shows bond lengths, bond angles, and energy differences of the each reaction. The reactions classified into two groups based on energy difference. Inter-dimer, intra-dimer, and inter-row were a group of favorable reactions and the others were a group of relatively unfavorable reactions. Bond lengths of Hf-O in favorable reactions were equally 1.93Å and bond lengths of O-O were shorter than the others. We compared the bond length of Hf-Cl and the bond angle of Cl-Hf-Cl with $HfCl_4$ molecule. The favorable reactions had relatively similar bond length of Hf-Cl and bond angle of Cl-Hf-Cl with $HfCl_4$ molecule. This result indicated that the configuration of $HfCl_2$ and oxygen atoms decided the stability of the reaction.

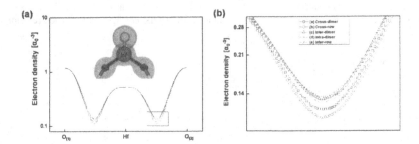

Figure 3. (a) The line profile of the electron density from $O_{(1)}$ to $O_{(2)}$; $O_{(1)}$-Hf- $O_{(2)}$ and (b) the expansion of square. The isosurface level was $0.08\ \alpha_0^{-3}$. $(\alpha_0$: Bohr radius)

Figure 4 shows the line profile of the electron density. The electron density was measured by the visualization for electronic and structural analysis (VESTA) and the isosurface level was $0.08\ \alpha_0^{-3}$ $(\alpha_0$: Bohr radius).[16] The electron densities of $O_{(1)}$-Hf and $O_{(2)}$-Hf were almost same. The square of figure 4 (a) was enlarged into figure 4 (b). Cross-dimer and cross-row reaction (classified unfavorable reactions) had lower electron densities of Hf-O than the others. It meant that the Hf-O bonds of cross-dimer and cross-row were weak, resulting in further energy increase.

CONCLUSIONS

The reaction of $HfCl_4$ with two hydroxyls on Si (001)-2×1 surface was investigated using DFT calculation. There were five reaction pathways of $HfCl_4$ for reacting with two hydroxyls. The reactions classified into two groups based on energy difference. The favorable reactions had short bond lengths of Hf-O and O-O than the others and they had similar bond length of Hf-Cl and bond angle of Cl-Hf-Cl with $HfCl_4$ molecule. The relatively unfavorable reactions had lower electron densities of Hf-O than the favorable reactions. These results indicated the configuration of $HfCl_2$ and oxygen atoms, the electron density of Hf-O decided the stability of the reaction.

REFERENCES

1. G. D. Wilk, R.M. Wallace, J. M. Anthony, *J. Appl. Phys.* **89**, 5243 (2001)
2. J. Robertson, *Rep. Prog. Phys.* **69**, 327 (2006)
3. K. Kukli, M. Ritala, J. Lu, A. harsta, and M. Leskela, *J. Electronchem. Soc.* **151**, 189 (2004)
4. P. D. Kirsch, M. A. Quevedo-Lopez, H. –J. Li, Y. Senzaki,, J. J. Peterson, S. C. Song, S. A. Krishnan, N. Moumen, J. Barnett, G. Bersuker, P. Y. Hung, B. H. Lee, T. Lafford, Q. Wang, D. Gay, and J. G. Ekerdt, *J. Appl. Phys.* **99**, 023508 (2006)
5. B. G. Willis, A. Mathew, L. S. Wielunski, and R. L. Opila, *J. Phys. Chem. C.* **112**, 1994 (2008)
6. C. Tang, B. Tuttle, and R. Ramprasad, *Phys. Rev. B.* 76, 073306 (2007)
7. G. Kresse and J. Hafner, *Phys. Rev. B.* **47**, 558 (1993); ibid. **49**, 14251 (1994).
8. G. Kresse and J. Furthüller, *Comput. Mat. Sci.* **6**, 15 (1996).

9. G. Kresse and J. Furthüller, *Phys. Rev. B.* **54**, 11169 (1996).
10. G. Kresse and D. Joubert, *Phys. Rev. B.* **59**, 1758 (1999).
11. D. Vanderbilt, *Phys. Rev. B.* **41**, R7892 (1990).
12. D. M. Wood and A. Zunger, *J. Phys. A.* **18**, 1343 (1985).
13. P. Pulay, *Chem. Phys. Lett.* **73**, 393 (1980).
14. H. -C. Oh, H. -I. Seo, and Y. -C. Kim, in *Applications of Group IV Semiconductor Nanostructures*, edited by T. van Buuren, L. Tsybeskov, S. Fukatsu, L. Dal Negro, F. Gourbilleau (Mater. Res. Soc. Symp. Proc. **1145E** Warrendale, PA, 2009), MM04-35.
15. D. -H. Kim, H. -I. Seo and Y. -C. Kim, *J. Kor. Vac. Soc.* **18**, 1 (2009)
16. K. Momma and F. Izumi, *J. Appl. Crystallogr.* **41**, 653 (2008).

Mater. Res. Soc. Symp. Proc. Vol. 1155 © 2009 Materials Research Society   1155-C01-05

Interaction of HfCl₄ precursor with H₂O terminated Si (001) surface: First principles study

Dae-Hyun Kim[1], Dae-Hee Kim[1], Hwa-Il Seo[2], and Yeong-Cheol Kim[1]
[1]Department of Materials Engineering, Korea University of Technology and Education, 1800 Chungjeollo, Byungchun-myun, Chonan, 330-708, Republic of Korea
[2]School of Information Technology, Korea University of Technology and Education, 1800 Chungjeollo, Byungchun-myun, Chonan, 330-708, Republic of Korea

ABSTRACT

We investigated the reaction of $HfCl_4$ molecules with a H_2O terminated Si (001)-2×1 surface using density functional theory to understand the initial stage of atomic layer deposition (ALD) of HfO_2. Half monolayer of H_2O molecules were adsorbed on the buckled-down Si atoms of the Si dimers of the Si (001)-2×1 surface below the dissociation temperature of H_2O and were dissociated into H and OH at room temperature. This process could make uniform and well-aligned -H and -OH's on the Si (001) substrate. The reaction of a $HfCl_4$ molecule was more favorable with -OH than -H. The reaction of the $HfCl_4$ molecule with the -OH generated a HCl molecule, and the remaining $HfCl_3$ was attached to the O atom. The first reaction of the $HfCl_4$ molecule with –OH produced 0.21 eV energy benefit. The reaction of the second $HfCl_4$ molecule with the most adjacent –OH of the first one produced 0.28 eV energy benefit. The third and fourth molecules showed same tendency with the first and second ones. The energy differences of the fifth and sixth $HfCl_4$ reactions were -0.01 eV, 0.06 eV, respectively. Therefore, we found that the saturation Hf coverage was approximately 5/8 of the available –OH's, which was 2.08 × 10^{14} Hf /cm². The result was well-matched with the experimental study of other group.

INTRODUCTION

The complementary metal oxide semiconductor (CMOS) device is the most important electronic device in microelectronic industry. Silicon dioxide (SiO_2) is used extensively as dielectric materials. However, SiO_2 is so thin that the tunneling leakage current becomes too high.[1] High-k materials have received much attention recently in microelectronics because high-k gate dielectrics can significantly suppress the tunneling leakage current in CMOS devices. Among various high-k materials, hafnium dioxide (HfO_2) is considered to be the most promising candidate due to its relatively high permittivity, good thermal stability, and compatibility with dual metal integration.[2] One of the best advantages for SiO_2 is that it can be grown by thermal oxidation; whereas, high-k oxides must be deposited.[2] Atomic layer deposition (ALD) technique is a desirable process to form HfO_2 because it shows good conformality and uniformity over large areas.[3,4]

Green et al. studied the efficacy of various underlayers for nucleation and growth of atomic layer deposited HfO_2 films. The use of a chemical oxide underlayer led to good HfO_2 nucleation with linear growth rate and almost no barrier. In contrast, the growth on H-terminated Si was characterized by a large barrier to nucleation and growth. It indicated that the initial preparation of the Si surface decided the film quality.[5] Willis et al. confirmed that a one-cycle ALD growth reaction on H_2O terminated Si (001)-2×1 surface was shown to lead successful nucleation, and the saturation Hf coverage was measured to be 2.08 × 10^{14} Hf /cm².[2, 6, 7]

In this study, we employed a first principles calculation to investigate the initial ALD mechanism of HfO_2 on H_2O-terminated Si (001) -2×1 surface. We used 4×4 Si surface that is covered with H's and OH's.

CALCULATION

Density functional theory (DFT) calculations were performed using the Vienna *ab-initio* Simulation Package (VASP) code with the projector augmented wave (PAW) potentials and the generalized gradient approximation (GGA).[8-12] The residual minimization scheme direct inversion in the iterative subspace (RMM-DIIS) was used for calculating the ground state of electrons.[13,14] A cutoff energy was 450 eV for the plane wave expansion of the wave functions and Monk-horst pack k-points mesh of 2×2×1 produced well converged results.

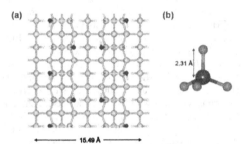

Figure 1. (a) A uniform H- and OH-terminated Si surface and (b) $HfCl_4$. Oxygen atoms are shown in red, hydrogen atoms in white, silicon substrate atoms in yellow, hafnium as the largest atoms in brown, and chlorines in green.

Figure 1 (a) shows a uniform H- and OH-terminated Si surface. This surface could be obtained through the 1/2 monolayer water molecule adsorption below the dissociation temperature of H_2O and dissociation above the dissociation temperature.[15] There are eight H's and OH's in this 4×4 Si (001) surface. Figure 2 (b) shows a $HfCl_4$ molecule. The distance between Cl and Hf atom was calculated to be 2.31 Å.

RESULTS AND DISCUSSION

The reactions of the $HfCl_4$ molecule with -H and -OH are shown in figure 2 (a). The reaction of the $HfCl_4$ molecule with -H was endothermic by 1.59 eV. In contrast, the reaction of the $HfCl_4$ molecule with -OH was exothermic by -0.17 eV. Therefore, the reaction of the $HfCl_4$ molecule was more favorable with -OH than -H. This result was well-matched with the experimental studies of other groups. When the $HfCl_4$ molecule was reacted with -OH, a HCl molecule was generated and the resulting $HfCl_3$ was attached to the O atom. The free energy could be influenced by the configuration of $HfCl_3$ with respect to the Si surface. Figure 2 (b) shows the energy difference as a function of the $HfCl_3$ angle. We obtained an optimal configuration of $HfCl_3$ at the angle of 180°.

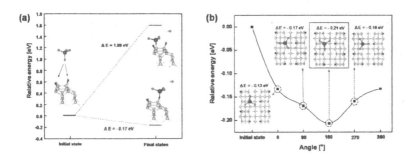

Figure 2. (a) The reactions of the HfCl$_4$ molecule with -H and -OH, and (b) the energy difference as a function of the HfCl$_3$ angle.

After the reaction of the first HfCl$_4$ with -OH, there were seven available -OH's on this 4x4 Si (001) surface for further reaction with other HfCl$_4$. We calculated the reaction of the next HfCl$_4$ molecules with OH's. Figure 3 shows the relative energy as a function of the number of the HfCl$_4$ molecules. The reaction of the second HfCl$_4$ molecule was favorable with the most adjacent -OH of the first one and it formed a pair with first one. The reaction of the third and fourth HfCl$_4$ molecule showed the same tendency with the first and second ones. The energy differences of the fifth and sixth HfCl$_4$ reactions were -0.01 eV, 0.06 eV, respectively. There were two points at this result. First, the energy differences of the second and fourth HfCl$_4$ reactions were a little higher than the first and third ones. Second, the reactions of HfCl$_4$ molecules were favorable up to fifth HfCl$_4$ and the reaction of sixth HfCl$_4$ molecule was endothermic by 0.06 eV.

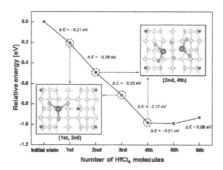

Figure 3. The relative energy as a function of the number of the HfCl$_4$ molecules.

Figure 4 shows the line profile of the electron density between two atoms. The electron density was measured by the visualization for electronic and structural analysis (VESTA) and the isosurface level was 0.08 α_0^{-3} (α_0: Bohr radius).[16] When the second HfCl$_4$ reacted with a -OH, it formed a pair with the first one, resulting in sharing Cl atoms. The electron density from Hf(1) to Cl(2) was lower than that from Cl(1) to Hf(1). It meant that the Hf atom shared the rest of electrons with the other Cl atom that was shared with Hf(2). It was confirmed that the electron density from Hf(1) to Cl(2) and from Hf(2) to Cl(2) were exactly same. Therefore, we believe that the sharing of Cl atoms from Hf atoms results in the further energy reduction.

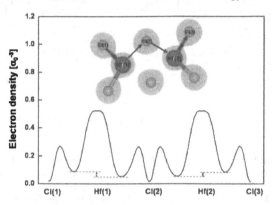

Figure 4. The line profile of the electron density from Cl(1) to Cl(3); Cl(1)-Hf(1)-Cl(2)-Hf(2)-Cl(3). The isosurface level was 0.08 α_0^{-3}. (α_0: Bohr radius)

Table 1. Distances and angles of HfCl$_3$, OH-terminated Si substrate, and the interface of HfCl$_3$-OH terminated Si substrate as a function of the number of HfCl$_4$.

	1st HfCl$_3$	2nd HfCl$_3$	3rd HfCl$_3$	4th HfCl$_3$	5th HfCl$_3$
Hf-Cl (Å)		2.56		2.56	2.37
	2.33	2.35	2.33	2.35	2.34
		2.34		2.33	2.31
∠Cl-Hf-Cl (°)	110.14	149.90	111.06	149.10	123.44
	109.68	97.51	109.88	97.27	98.59
	109.51	87.91	109.77	88.49	96.91
Si-Si	2.44	2.43	2.46	2.45	2.48
Si-O	1.69	1.69	1.69	1.69	1.67
∠Si-Si-O	108.58	108.85	108.53	109.47	119.65
Hf-O	1.90	1.89	1.90	1.89	1.93
∠Si-O-Hf	164.88	166.57	165.51	165.46	173.02
∠O-Hf-Cl	111.72	106.77	110.52	106.86	114.56
	110.34	104.95	109.46	104.37	114.01
	105.38	100.26	106.02	100.95	103.24

68

The configurations as a function of the number of the $HfCl_4$ molecules were shown in table 1. It consists of three parts; the $HfCl_3$, OH-terminated Si, and the interface of $HfCl_3$-OH terminated Si substrate. The first $HfCl_3$ showed almost same configuration with the third one and the second $HfCl_3$ showed almost same configuration with the fourth one. There was no significant change of the substrate and the interface up to the reaction of the fourth $HfCl_4$. However, the reaction of the fifth $HfCl_4$ caused lots of change in $HfCl_3$, the substrate, and the interface compared to previous ones. Two pairs of $HfCl_3$'s (the first one and the second one, the third one and the fourth one) were so favorable that the fifth $HfCl_3$ suffered the repulsion force caused by the previous four $HfCl_3$'s due to the steric hindrance, resulting in significantly reduced energy reduction. The endothermic reaction of the sixth $HfCl_4$ was explained by the same reasoning. In the 4×4 Si surface size, therefore, the reactions of $HfCl_4$ molecules were favorable up to fifth $HfCl_4$. It demonstrated that the saturation Hf coverage was calculated to be 2.08×10^{14} Hf/cm^2 and this result was well-matched with the experimental data of other group.

CONCLUSIONS

We performed a density functional theory study to investigate the reaction of $HfCl_4$ with H_2O-terminated Si (001)-2×1 surface. The reaction of the $HfCl_4$ molecule was more favorable with -OH than -H. When even numbers of the $HfCl_4$ molecules reacted –OH's on the surface, the molecules shared some Cl atoms, resulting in further energy reduction. In 4×4 Si surface size, the reactions of $HfCl_4$ molecules were favorable up to fifth $HfCl_4$. The calculated saturation Hf coverage was 2.08×10^{14} Hf/cm^2 and was well-matched with the experimental data.

REFERENCES

1. G. D. Wilk, R.M. Wallace, J. M. Anthony, *J. Appl. Phys.* **89**, 5243 (2001)
2. J. Robertson, *Rep. Prog. Phys.* **69**, 327 (2006)
3. K. Kukli, M. Ritala, J. Lu, A. harsta, and M. Leskela, *J. Electronchem. Soc.* **151**, 189 (2004)
4. P. D. Kirsch, M. A. Quevedo-Lopez, H. -J. Li, Y. Senzaki,, J. J. Peterson, S. C. Song, S. A. Krishnan, N. Moumen, J. Barnett, G. Bersuker, P. Y. Hung, B. H. Lee, T. Lafford, Q. Wang, D. Gay, and J. G. Ekerdt, *J. Appl. Phys.* **99**, 023508 (2006)
5. M. L. Green, M. -Y. Ho, B. Busch, G. D. Wilk, T. Sorsch, T. Conard, B. Brijs, W. Vandervorst, P. I. Raisanen, D. Muller, M. Bude, and J. Grazul, *J. Appl. Phys.* **92**, 7168 (2002)
6. A. Mathew, L. S. Wielunski, R. L. Opila, and B. G. Wills, *ECS Transactions.* **11**, 183 (2007)
7. B. G. Willis, A. Mathew, L. S. Wielunski, and R. L. Opila, *J. Phys. Chem. C.* **112**, 1994 (2008)
8. G. Kresse and J. Hafner, *Phys. Rev. B.* **47**, 558 (1993); ibid. **49**, 14251 (1994).
9. G. Kresse and J. Furthüller, *Comput. Mat. Sci.* **6**, 15 (1996).
10. G. Kresse and J. Furthüller, *Phys. Rev. B.* **54**, 11169 (1996).
11. G. Kresse and D. Joubert, *Phys. Rev. B.* **59**, 1758 (1999).
12. D. Vanderbilt, *Phys. Rev. B.* **41**, R7892 (1990).
13. D. M. Wood and A. Zunger, *J. Phys. A.* **18**, 1343 (1985).
14. P. Pulay, *Chem. Phys. Lett.* **73**, 393 (1980).
15. H. -C. Oh, H. -I. Seo, and Y. -C. Kim, *2008 MRS fall meeting* (2008)
16. K. Momma and F. Izumi, *J. Appl. Crystallogr.* **41**, 653 (2008).

Mater. Res. Soc. Symp. Proc. Vol. 1155 © 2009 Materials Research Society   1155-C11-02

Hafnia Surface and High-k Gate Stacks

X. Luo[1], Alexander A. Demkov[1], O. Sharia[1], G. Bersuker[2]

[1]Department of Physics, The University of Texas at Austin, Austin, Texas 78712, USA
[2] SEMATECH, Austin, Texas 78741, USA

ABSTRACT

Hafnium dioxide that belongs to a class of metal oxides with a high dielectric constant or high-k dielectrics has been recently introduced as a gate dielectric in field effect transistors. We report a theoretical study of structural and electronic properties of hafnia surface, and the electronic structure and band alignment at hafnia interfaces with metals, oxides and semiconductors that are crucial in gate stack engineering.

INTRODUCTION

As scaling of the complementary metal oxide semiconductor (CMOS) technology takes us below 45 nm many new materials, traditionally not associated with the semiconductor process, are being introduced into manufacturing. Notably, several transition metal (TM) oxides, or more generally dielectrics with a high dielectric constant or high-k dielectrics, are being considered for the gate stack applications instead of SiO_2. The gate stack is a multilayer structure in place of the metal oxide semiconductor capacitor. Its capacitance controls the transistor saturation current and has been traditionally maintained by reducing its thickness in accord with the gate length reduction (the so-called scaling at the heart of Moor's law). However, after reaching the oxide thickness of 12 Å the scaling has stopped due to the prohibitively large gate leakage current caused by direct tunneling across the gate oxide. Thus a new dielectric with a larger dielectric constant had to be introduced [1].

In the current generation of field-effect transistors (FETs) hafnia-based dielectric films started to replace silica as a gate dielectric [1,2]. Hafnia (HfO_2) has been selected due to its high thermodynamic stability on Si [3,4], sufficiently wide band gap, and high dielectric constant (k=20-25) [5]. Bulk hafnia crystallizes in three polymorphs; at room temperature and ambient pressure it is monoclinic, and transforms first to a tetragonal form and then to a fluorite-type cubic form at elevated temperatures. Thin hafnia films are typically amorphous as deposited, and polycrystalline after a post-deposition anneal, typically they are mixed phase [6] with mainly monoclinic $(\bar{1}11)$, (001) and (111) textures [7,8,9]. The structure-property relation of thin hafnia films is not completely clear. In recent years the bulk properties of hafnia, such as structure, electronic spectrum and dielectric constant have been studied theoretically [5]. However, thin hafnia films behave differently from bulk hafnia. In particular, phase transitions occur at much lower temperature. In addition, the surface of hafnia may play a role in the gate dielectric film thermodynamics due to film's high surface area to volume ratio. For example, a low surface energy phase may be stabilized below a critical thickness. Thus a systematic study of hafnia surface including monoclinic and tetragonal phases for both stoichiometric (the surface composition is HfO_2) and non-stoichiometric terminations is of interest. Recently, Iskandarova [10] and Mukhopadhyay [11] reported theoretical studies of stoichiometric surfaces of the monoclinic phase of hafnia. We will review some of our recent theoretical studies of non-stoichiometric hafnia surfaces [12].

In the case of hafnia films grown on a Si substrate a thin SiO_2 layer, grown either intentionally or spontaneously during hafnia deposition and subsequent fabrication processing, is always present at the interface between the high-k film and Si substrate. Using density functional theory we show that the band offset between SiO_2 and HfO_2 clearly determines the overall alignment of the gate stack [13]. Results of the first principles calculations are in good agreement with recent experiment [14].

COMPUTATIONAL DETAILS

Most of our theoretical calculations are done within the generalized gradient approximation (GGA) to density functional theory as incorporated in the VASP code [15]. When the LDA is used we employ the Ceperley-Alder exchange-correlation functional [16] as interpolated by Perdew and Zunger [17]. We use the Vanderbilt-type ultra-soft pseudopotentials [18], and a conventional plane wave basis set. We use projector-augmented wave (PAW) method to study the site projected density of states [19]. For the Brillouin zone integration we use the Monkhorst-Pack [20] special k-point meshes. The total energy is converged to 0.005 eV/cell. Atomic positions are relaxed with the conjugate gradient method with force tolerance of 0.05eV/Å. To simulate interfaces we use super cell geometry with both vacuum slabs and superlattices.

DISCUSSION

Hafnia surface

In Fig. 1 we plot the surface energy as function of the oxygen chemical potential for different orientations and stoichiometry for monoclinic HfO_2. Assuming the surface is in equilibrium with the bulk: $\mu_{Hf} + \mu_{O_2} = -E_{HfO_2}$, the surface free energy can be expressed as function of one variable. We choose the chemical potential of oxygen measured with respect to O_2 molecule. The zero value of chemical potential indicates equilibrium with the oxygen supply and thus describes the oxygen rich environment. As seen in Fig. 1 all stoichiometric surfaces have surface energies independent of chemical potential. For the stoichiometric surfaces our calculations agree well with those presented in references 10 and 11. In addition, we consider several non-stoichiometric and H-passivated surfaces as shown in Figure 1. Two surfaces with the lowest surface energy among the non H-passivated surfaces are the stoichiometric $(11\bar{1})$ and oxygen rich $(11\bar{2})$. This might explain why thin hafnia films grown by atomic layer deposition [12] favor the texture axis normal to $(11\bar{1})$ and $(11\bar{2})$ planes. However, as seen in Fig. 1, under certain conditions hafnia films favor (001) as the growth direction. Our calculations also suggest that in the presence of hydrogen the $(11\bar{2})$ termination can be stabilized even under oxygen poor conditions, which might explain the peculiar texturing of very thin monoclinic films [12]. Our results suggest that thermodynamics plays a role in hafnia films growth.

Though previous studies [6-11] didn't mention the monoclinic $(11\bar{2})$ surface, our experimental [12] identify this surface as thermodynamically stabile. Theoretically we find that two kinds of peroxy bonds form on this surface, resulting in the electronic state in the band gap. One is an anti-bonding $pp\pi$ orbital of the peroxy dimer and the other is an anti-bonding $pp\sigma$ orbital. It will be of interest to identify these bonds experimentally. In Figure 2 we show the phonon density of states for the bulk monoclinic phase and $(11\overline{2})$ surface. The frequencies of the peroxy bond vibrations are higher than those of crystalline phonons. This could be used as a finger print identifying peroxy bonds in e.g., infra red optical experimental studies.

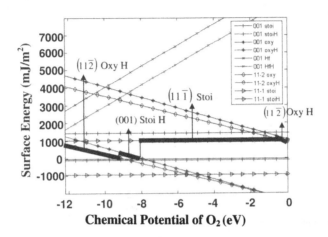

Figure 1. Surface energy of several surfaces of monoclinic HfO_2 as function of the oxygen chemical potential. Labels ending in "Oxy", "Hf" and "Stoi" refer to oxygen, hafnia, and stoichiometric terminations, respectively. Labels ending in 'H' refer to hydrogen passivation. Thick lines indicate the stable surfaces under certain ranger of chemical environment.

In addition, we have examined the work function and electron affinity for several surfaces of monoclinic and tetragonal hafnia. For the monoclinic phase the electron affinity shows little variation with the crystalline orientation or termination and is about 1.7 eV for the most stable (111) oriented stoichiometric surface in good agreement with experiment. In the case of tetragonal HfO_2 we find a large variation in electron affinity with surface termination. For the oxygen-terminated (111) surface the electron affinity is estimated to be 2.52 eV when a quisiparticle correction is included. Fig.3 shows a combined plot of the plane-averaged electrostatic potential and its macroscopic average for hafnia-terminated (111) slab along with the plane by plane projected oxygen density of states. Oxygen is chosen since the top of the valence band in hafnia is predominantly the oxygen p-state. The Fermi level is pinned by the mid-gap surface state. The state is mostly a dangling hafnia d-orbital but is hybridized with the oxygen p-orbital. We also estimate the electron affinity using the bulk value of the reference potential. In the bulk the valence band top is 2.4 eV higher than the average electrostatic potential. Using the experimental value of the HfO_2 band gap of 5.7 eV we estimate the electron affinity to be 1.95 eV. Applying a GW correction [21] of 0.57 eV (calculated for bulk m-HfO_2 [22]) we arrive at the value of 2.52 eV similar to the recent experimental result, 2.5 eV [14].

Hafnia-silica interface

Experiment shows significant variation in the band offset at the SiO_2-HfO_2 interface. The study of the HfO_2-SiO_2-S+i gate stack was performed by Sayan, Emge, Garfunkel and co-workers using

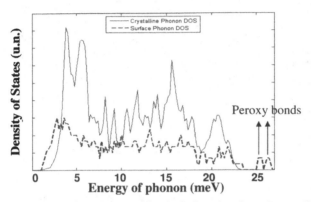

Figure 2. Density of phonon states for the bulk and surface of monoclinic phase hafnia. The solid line refers to the crystalline phonons and the dash line refers to the surface phonons. The two arrows show the energy position of two peroxy bonds.

a combination of x-ray and inverse photoemission and *ab-initio* theory [23]. From Sayan *et al.* we infer the SiO_2/HfO_2 valence band offset to be between 0.89 and 1.25 eV depending on the method of analysis. To understand this behavior let us analyze the mechanism of band alignment. Before the oxides are brought into contact the band discontinuity is given by the Schottky rule. However, when the oxides are brought together the charge transfer becomes possible, and a correction needs to be added. With respect to vacuum, the top of the valence band in hafnia is at higher energy than that in silica before the contact, so the charge transfer would be from hafnia to silica. The valence electron density should undergo a smooth transition from the hafnia value of about 0.5 e/$Å^3$ to that in silica (0.3 e/$Å^3$) as required by the kinetic energy term in the Hamiltonian. In the case of a metal surface Smoluchowsky called this effect "spreading" [24]. Alternatively, the interface dipole can be seen as associated with the difference in the charge neutrality levels (CNL) of the two insulators. "Locally" the charge transfer may be thought of in terms of the difference in the electro-negativity of metals (1.3 and 1.9 for Hf and Si, respectively) which ultimately defines the CNLs. Although the VBTs of silica and hafnia are predominantly oxygen levels, the coordination of oxygen and its hybridization with the metal are somewhat different in two materials. Regardless of the specific model a dipole layer would form across the 2-4 Å of the interface (compare with d_{SiO}+d_{HfO}=3.6 Å) and shift the Schottky answer. Simple chemistry dictates that oxygen bridges connect HfO_2 to SiO_2. The induced polarization of this oxygen reduces the internal field in the dipole layer caused by the "spreading", and moves the offset back to the Schottky limit. Importantly, we find that the dielectric constant of this "dielectric" is a function of the average oxygen coordination. For the total band offset we have [13]:

$$V_{vb} = V_{schottky} - \frac{\bar{\rho}d^2}{2\varepsilon_0\varepsilon} = 1.6 - \frac{13.5}{\varepsilon} \ (eV) \tag{1}$$

Here ε is the dielectric response of the interfacial layer, and d is the thickness of the interface layer (we take $d = 1.4\text{Å}$, approximately the distance between two atomic planes), $\bar{\rho}$ is a half of the difference between two densities or 0.076 electrons/Å^3.

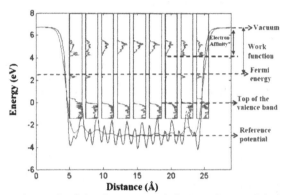

Figure 3. A composite graph of the plane averaged electrostatic potential used as the energy reference, its macroscopic average, and the plane by plane projected oxygen density of states for the Hf-terminated (111) surface of t-HfO$_2$. The surface state which is actually a dangling Hf d-orbital is clearly seen pinning the Fermi level.

We find the dielectric constant is a linear function of the oxygen coordination backed out from the *ab-initio* result using this expression. It varies smoothly form the silica-like to hafnia-like value (4 and 22, respectively) as we go from two-fold to three-fold interface. Considering that the electronic component is small (2 for silica [25], and 5 for hafnia [26]) we attribute the coordination dependence of the dielectric constant to the lattice polarizability. The latter can depend on the local geometry (bonding) either through the vibrational mode frequency ω_λ or through the Born effective charge $\tilde{Z}_{\lambda a}^*$. We assign the dependence to the Born effective charge. Assuming the average ε_∞ of 3.5 and fitting the value of z^* for $N_{average} = 3$ we use a crude approximation $\varepsilon_{lattice} = \beta(Z_{N_{average}}^*)^2$ and extract the oxygen Born effective charge as a function of its average coordination. The experimental values of the band discontinuity therefore indicate a reasonably good interface with the average oxygen coordination of 2.5. The dependence of the discontinuity on the oxygen coordination suggests a way of tuning the alignment across the gate stack. Indeed, our calculations suggest that when the stack is doped with Al, Al tends to accumulate near the SiO$_2$/HfO$_2$ interface thus reducing the amount of oxygen in that region [27]. Since this oxygen is responsible for screening the interfacial dipoles formed due to the charge transfer, higher oxygen deficiency of the interfacial region leads to an increase of the effect of the interface dipole on band alignment shifting it away from the Schottky limit.

CONCLUSION

We report theoretical studies of hafnia surfaces and band alignment in hafnia gate stacks. The calculated surface phase diagram of hafnia explains why hafnia films choose certain texturing under particular chemical conditions. The results indicate that thermodynamics plays an important role during the film growth. We find that Al in the SiO_2 layer near its interface with HfO_2 changes the band discontinuity significantly, thus offering a way to adjust the overall band alignment in the stack.

ACKNOWLEDGEMENTS

This work in part is supported by the National Science Foundation under grants DMR-0548182, DMR-0606464, the Welch Foundation under grant F-1624, and Texas Advanced Computing Center. We thank John Robertson for many insightful discussions.

REFERENCE

1. G.D. Wilk, R.M. Wallace, and J.M. Anthony, *J. Appl. Phys.*, **89**, 5243 (2001).
2. 2007 International Technology Roadmap for Semiconductors, Semiconductor Industry Association, San Jose (2007).
3. M. Copel, M. Gribelyuk, and E. Gusev, *Appl. Phys. Lett.* **76**, 436 (2000).
4. M. Gutowski, J. E. Jaffe, C. Liu, M. Stoker, R. I. Hegde, R.S. Rai and P. J. Tobin, *Appl. Phys. Lett.*, **80**, 1897, (2002).
5. X. Zhao and D. Vanderbilt, *Phys. Rev.* B **65**, 233106 (2002).
6. P. Lysaght, J.C. Woicik, B.H. Lee, R. Jammy, *Appl. Phys. Lett.* **91**, 122910, (2007).
7. M. Y. Ho, H. Gong, G. D. Wilk, B. W. Busch, M. L. Green, D. A. Muller, M. Bude, W. H. Lin, M. E.Loomans, S. K. Lahiri, and P. T. Räisänen, *J. Appl. Phys.* **93**, 1477 (2003).
8. D. Triyoso, R. Liu, D. Roan, M. Ramon, N. V. Edwards, R. Gregory, D. Werho, J. Kulik, G. Tam, E. Irwin, X. D. Wang, L. B. La, C. Hobbs, R. Garcia, J. Baker, B. E. White, and P. Tobin, *J. Electrochem. Soc.* **151**, 220 (2004).
9. J. Aarik, A. Aidla, V. Sammelselg, and T. Uustare, *J. Cryst. Growth* **220**, 105 (2000).
10. I.M. Iskandarova, A.A. Knizhnik, E.A. Rykova, A.A. Bagatur'yants, B.V. Potaptkin, and A.A. Korkin, *Microelectronic Eng.* **69**, 587 (2003).
11. A.B. Mukhopadhyay, J.F. Sanz, and C.B. Musgrave, *Phys. Rev.* B **73**, 115330 (2006).
12. X. Luo, A.A. Demkov, D. Triyoso, P. Fejes, R. Gregory, and S. Zollner, *Phys. Rev.* B **78**, 245314 (2008).
13. O. Sharia, A.A. Demkov, G. Bersuker, and B.H. Lee, *Phys. Rev.* B **75**, 035306 (2007).
14. E. Bersch, S. Rangan, R.A. Bartynsky, E. Gaefunkel, and E. Vescovo, *Phys. Rev.* B **78**, 085114 (2008).
15. G. Kresse and J. Furthmuller, *Phys. Rev.* B **54**, 11169 (1996).
16. D. M. Ceperley and B. J. Alder, *Phys. Rev. Lett.* **45**, 566 (1980).
17. J. P. Perdew and A. Zunger, *Phys. Rev.* B **23**, 5048, (1981).
18. P. E. Blöchl, *Phys. Rev.* B **50**, 17 953 (1994).
19. D. Vanderbilt, *Phys. Rev.* B **41**, 7892 (1990).
20. H. J. Monkhorst and J. D. Pack, *Phys. Rev.* B **13**, 5188 (1976).
21. L. Hedin, *Phys. Rev.* **139**, A796 (1965).

22. O. Sharia, *private communication*.
23. S. Sayan, T. Emge, E. Garfunkel, X. Zhao, L. Wielunski, R.A. Bartynski, D. Vanderbilt, J.S. Suehle, S. Suzer, and M. Banszak-Holl, *J. Appl. Phys.* **96**, 7485 (2004).
24. R. Smoluchowsky, *Phys. Rev.* **60**, 661 (1941).
25. G.-M. Rignanese, X. Gonze, K. Cho, A. Pasquarello, *Phys. Rev.* B **69**. 184301 (2004)
26. X. Gonze, D.C. Allan and M.P. Teter, *Phys. Rev. Lett.* **68**, 3603 (1992).
27. O. Sharia, A.A. Demkov, G. Bersuker, and B.H. Lee, *Phys. Rev.* B **77**, 085326 (2008).

Mater. Res. Soc. Symp. Proc. Vol. 1155 © 2009 Materials Research Society   1155-C05-03

Ultimately Low Schottky Barrier Height at NiSi/Si Junction by Sulfur Implantation After Silicidation for Aggressive Scaling of MOSFETs

Yen-Chu Yang[1], Yoshifumi Nishi[2], and Atsuhiro Kinoshita[2]
[1]Department of Materials Science and Engineering, Stanford University, CA, USA
[2]Advanced LSI Technology Laboratory, Corporate R&D Center, Toshiba Corporation, Yokohama, Japan

ABSTRACT

We have obtained the Schottky barrier height (SBH) of 3.4meV for an electron at the interface of nickel-silicide (NiSi) and silicon (Si) interface using sulfur (S) implantation after silicidation (S-IAS) process followed by drive-in annealing process. This value of SBH is much smaller than those previously reported. The NiSi/Si interface morphology observed by TEM indicating well interfacial flatness and no degradation occurs by sulfur implantation. The secondary-ion-mass-spectroscopy (SIMS) analysis result showed that S diffusion was suppressed by S-IAS process, leading to such a small value of SBH. In addition, S-IAS process was applied to 50nm MOSFET and the parasitic resistance was effectively lowered.

INTRODUCTION

Parasitic resistance, particularly source/drain (S/D) contact resistance becomes one of the most serious problems to extend MOSFET scaling recently. To lower the contact resistance, Schottky barrier at the interface of S/D metal and Si should be lowered. Nickel silicide (NiSi), with advantages of low resistivity and high scalability, has been used as the S/D metal. However, its Schottky barrier height (SBH) of 0.65eV for electrons is too high to achieve the contact resistance necessary for further scaling especially beyond 16nm generations.

One of the main streams for SBH reduction is replacing NiSi with another silicide metal [1,2]. Recently, new techniques such as Ni alloy silicide [3], metal segregation [4] and so on have been proposed. However, applying these techniques still needs much cost and time for commercial use. Among several solutions, high concentration impurity-segregation layers have been introduced to reduce SBH [5,6]. Sulfur (S) has been considered to be an efficient material to reduce SBH by removing Fermi-pinning [7,8]. Previous studies have investigated the segregation by implanting S before NiSi formation [9]. However, the high diffusivity of S in Si leads to loss of S concentration at the interface during thermal treatment [10]. Moreover, S ions spread into the substrate and channel region generate deep impurity levels that induce junction leakage and off leakage, resulting in device performance degradation. In this paper, S implantation after silicidation (S-IAS) process is proposed to suppress S diffusion.

NiSi/Si Schottky diodes were fabricated on an *n*-type and *p*-type Si(100) wafer. The thickness of NiSi was 16nm. Small amount of Pt was added to Ni to enhance the thermal stability of the silicide. S was implanted into NiSi/Si diodes at the energy of 10keV with doses of 5×10^{14} and 1×10^{15} cm^{-2}. The projection range of S in NiSi was about 6nm. Drive-in annealing was performed at 300°C and 450°C to compare the effect of heat treatment. Transmission electron microscope (TEM) was used to observe the morphology of NiSi/Si interface after S implantation. The doping profiles of S were analyzed by secondary-ion mass spectroscopy (SIMS). The Schottky diodes were characterized by current-voltage (*I-V*) measurement at various temperatures to extract the SBH. Moreover MOSFETs with NiSi electrodes with S implantation were also fabricated to study the effect of S segregation on parasitic resistance.

RESULT

The cross-sectional TEM image of the NiSi/*n*-Si diode with S implantation at 10keV and at a dose of 1×10^{15} cm^{-2} followed by drive-in anneal at 450°C for 1 min is shown in Fig.1. It is clear that NiSi layer holds high uniformity after S implantation. Namely, no morphology degradation occurs through S-IAS process.

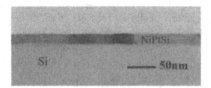

Figure 1. The cross-sectional TEM image of NiSi/Si diode with S implantation at 10keV and 1×10^{15}cm^{-3} followed by drive-in anneal at 450°C for 1 min.

The S profile of the same diode measured by backside SIMS is shown in Fig.2 (a). The NiSi/Si interface is located at the depth of 16nm, which can be determined by Ni profile (not shown). Although the S profile on the NiSi side is not qualitatively accurate, it can be seen from the profile on the Si side that segregation of S occurs at the interface with the concentration of higher than 4×10^{20} cm^{-3}. It should be noted that the abruptness of S inside Si is much sharper than that reported in the literature [9], which supports the hypothesis that S diffusion and the loss of S concentration at the interface are suppressed through our S-IAS process. The obtained S profile and ratio of the concentration to that of the NiSi/Si diode without drive-in annealing process are shown in Fig.2 (a) and (b), respectively. It can be seen that S concentration at the NiSi/Si interface increases with the drive-in annealing, and instead, that inside of the NiSi near the interface decreases. This indicates that S is energetically favorable to accumulate at the interface of NiSi through the thermal treatment.

The *I-V* characteristics at room temperature for NiSi/Si diodes with S dose of 1×10^{15}cm^{-2} under different drive-in annealing temperatures are shown in Fig.3. The reverse current for the diode on the *n*-Si substrate becomes larger as the annealing temperature becomes larger, indicating that the SBH for an electron becomes lower with the temperature. This is consistent with the fact the reverse current on the *p*-Si becomes smaller with the temperature, indicating that the SBH for a hole becomes higher.

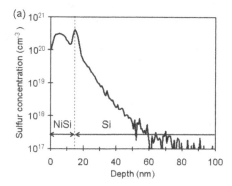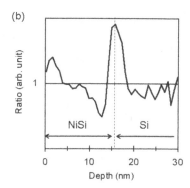

Figure 2. (a) The SIMS image of S profile that S was implanted into NiSi/Si diode at 10keV with the dose of $1 \times 10^{15} cm^{-2}$, followed by drive-in annealing at 450C for 1min. (b) The ratio of S concentration in the NiSi/Si diodes with and without drive-in annealing process.

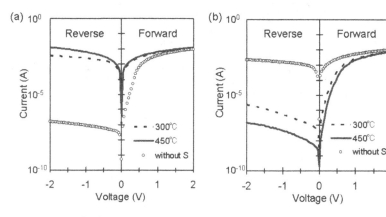

Figure 3. *I-V* characteristic of (a) NiSi/*n*-Si diodes and (b) NiSi/*p*-Si with S-implantation at various annealing temperatures with a dose of $10^{15} cm^{-2}$.

The SBH was extracted with the *I-V* characteristics at various temperatures, from 45K to 80K, based on the assumption that thermionic emission is dominant. The SBH for an electron under various experimental conditions are listed in Table 1. The extracted SBH values are comparable to the temperatures of the measurement; however, the obtained values of the ideal factor *n* between 1.01 and 1.08 support that the thermionic emission theory is still applicable to our experiment. The SBH of NiSi/*n*-Si decreases with increasing S implantation dose and annealing temperature. The SBH of NiSi/Si is 650meV without S-implantation. The literature-reported lowest SBH was 70meV with S implantation before Ni silicidation [14]. With using S-

IAS process, we have successfully obtained the SBH as low as 3.4meV with the dose of 10^{15}cm^{-2} on the drive-in annealing at 450°C. Since the S-IAS condition has not totally optimized yet, a further SBH reduction may be realized. It should be noted that the obtained SBH of 3.4meV is much lower than the thermal energy at room temperature, 27meV. Therefore, NiSi/n-Si junction with S doping through S-IAS process could be work as the ideal Ohmic contact at room temperature.

S dose (cm^{-2})	Annealing temperature	SBH (meV)
1x10^{15}	300°C	27
5x10^{14}	300°C	36
1x10^{15}	450°C	3.4
5x10^{14}	450°C	5.5

Table 1. SBH value of NiSi/n-Si diodes under different doping and annealing conditions.

The I_d-V_d curves of the n-type MOSFETs with and without S-implantation after NiSi formation are demonstrated in Fig.4. Figure 4 (a) shows the I_d-V_d curves for the S-implanted MOSFET with dose of 1×15 cm^{-2} and annealing at 450C whereas Fig.4 (b) shows those without S-implantation. Both MOSFETs were designed with a gate length of 50nm. It is clear that the MOSFET with S-implantation has higher on current than the one without S-implantation. The larger slope at small V$_d$ in Fig.4 (a) reveals the fact that the contact resistance, which is a main source of parasitic resistance, is smaller for the S-implanted MOSFET. This provides an effective way to realize extreme low parasitic resistance.

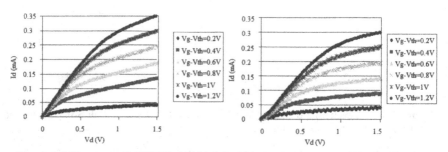

Figure 4. I_d-V_d curves of MOSFETs (a)with S-implantation and (b)without S-implantation.

CONCLUSION

In this work, S-IAS process was used to incorporate S atoms at the interface of NiSi/Si in order to reduce the Schottky Barrier Height (SBH) for an electron. S-IAS process successfully suppresses the S diffusion and loss in the concentration at the interface. The I-V characteristics show that SBH is sufficiently shrunk as the annealing temperature becomes higher, reaching the value as low as 3.4meV, the lowest value ever reported, for the diode where S was implanted at 10keV and $1 \times 10^{15} cm^{-2}$ followed by the drive-in annealing at 450°C. The SIMS analysis result supports the hypothesis that the high S concentration is obtained with the S-IAS process. Moreover, despite the worry that S-implantation might damage the NiSi/Si interface morphology, cross sectional TEM images show that the interfacial flatness was not damaged, indicating that no degradation occurs by S implantation. The MOSFET operation demonstrated that the parasitic resistance was effectively lowered by S implantation. In summary, S-IAS process, which provides ultimately low SBH at NiSi/Si interface, is a promising technique to realize ultra-low parasitic resistance source/drain for future LSI beyond 16nm generation.

REFERENCES

1. Shiyang Zhu, et al., "N-Type Schottky Barrier Source/Drain MOSFET Using Ytterbium Silicide", *IEEE Electron Device Letters*, vol.25, no.8, pp.565-567, 2004.
2. L.E. Calvet, H. Luebben, M.A. Reed, C. Wang and J.P. Snyder, "Suppression of leakage current in Schottky barrier metal-oxide-semiconductor field-effect transistors", J. Appl. Phys., **91**, 757 (2002)..
3. Rinus T.P. Lee, et al., "Novel Nickel-Alloy Silicides for Source/Drain Contact Resistance Reduction in N-Channel Multiple-Gate Transistors with Sub-35nm Gate Length", *IEDM Tech. Dig. 2006*, p.851.
4. Y. Nishi, Y. Tsuchiya, A. Kinoshita, T Yamauchi and J. Koga, "Interfacial Segregation of Metal at NiSi/Si Junction for Novel Dual Silicide Technology," *IEDM Tech. Dig. 2007*, p.135.
5. A. Kinoshita, Y. Tsuchiya, A.Yagishita, K. Uchida and J. Koga, "Solution for High-Performance Schottky-Source/Drain MOSFETs: Schottky Barrier Height Engineering with Dopant Segregation Technique", *2004 Symp. VLSI Tech.*, p.168.
6. T. Yamauchi, Y. Nishi, Y. Tsuchiya, A. Kinoshita, J. Kog and K. Kato, "Novel doping technology for a 1nm NiSi/Si junction with dipoles comforting Schottky (DCS) barrier," *IEDM Tech. Dig. 2007*, p.963.
7. R. Saiz-Pardo, R. Perez, F. J. Garcia-Vidal, R. Whittle, and F. Flores, *Surf. Sci.* 426, **26**, 1999.
8. K. Ikeda, Y. Yamashita, N. Sugiyama, N. Taoka, and S. Takagi, "Modulation of NiGe/Ge Schottky barreir height by sulfur segregation during Ni germanidation," *Appl. Phys. Lett.*, **88**, 152115, 2006
9. Q. T. Zhao, U. Breuer, E Rije, St. Lenk, and S. Mantl, "Tuning of NiSi/Si Schottky barrier heights by sulfur segregation during Ni silicidation," *Appl. Phys. Lett.*, **86**, 062108, 2005.
10. .H.-S. Wong, "Selenium Co-implantation and Segregation as a New Contact Technology for Nanoscale SOI N-FETs Featuring NiSi:C formed on Silicon-Carbon (Si:C) Source/Drain Stressors", *2008 Symp. VLSI Tech.* p.168.

Mater. Res. Soc. Symp. Proc. Vol. 1155 © 2009 Materials Research Society   1155-C05-06

Ammar M. Nayfeh and Viktor I. Koldyaev

Innovative Silicon, 4800 Great America Parkway, Suite 500 Santa Clara, CA 95054
Phone: 408-969-9555 x2043 Fax 408-969-2367, Email: anayfeh@z-ram.com

ABSTRACT

Point Defect (PD) mediated diffusion of phosphorous in silicon is studied in order to address the long standing open problem of PD-Dopant pair lifetime. A novel experimental method is suggested to increase PD-P pair lifetime for better observability and experimental resolution. In the experiment, phosphorous is implanted, followed by low temperature poly-Si deposition with in-situ doped phosphorous. The P profile shows, after low temperature ($<650°C$) in-situ phosphorous doped poly-Si deposition, an exponential dependence of two orders of magnitude for a significant depth scale. This indicates that the PD-P pairs survive long-range diffusion before dissociating in the Si lattice. As a result, the lifetime of PD-P pair was extracted and this provides a physical basis for TCAD simulation at the atomic scale.

INTRODUCTION

Point defect mediated dopant diffusion is established as the microscopic mechanism for diffusion of dopants in semiconductors [1-3]. A silicon interstitial (Si_I) is temporarily bound with a dopant atom while the pair migrates as a mobile species before it dissociates leaving the dopant atom on a substitutional site and releasing the silicon interstitial. However, experimental observation of this diffusion is hampered due to low lifetime of pairs and hence its quantification is a long standing open problem. Cowern et al. [2] showed exponential Boron (B) profile confirming B-Si_I pair diffusion. In this work, a novel experiment is carried out to study Phosphorus (P) Interstitial pair (P-Si_I) diffusion. The experiment resulted in observability of P-Si_I pair diffusion for significant depth scale. Solution to the Boundary Value Problem (BVP) allowed for extraction of lifetime and diffusivity of P-Si_I pairs. Pair lifetime quantification is important for more accurate Physics Based (PB) TCAD simulation in modern processing condition [4].

EXPERIMENTAL THEORY

In order to observe pair diffusion novel experimental methods are needed. Low temperature processing is required for better pair lifetime characterization while excess I generation is a prerequisite for P-Si_I pair diffusion observability. P drive-in from solid state source has been shown to generate I at super-saturation level [1]. In addition, a shallow low dose P implant provides additional Si_I and can be used as a marker layer for P diffusion observation. The effect of interstitials created by P implant is negligibly small because of two features of the experiment. For a low P dose and high processing T (T>550°C), Transient Enhanced Diffusion (TED) due to interstitials created by P implantation decays significantly faster than the poly-Si deposition time (annealing for interstitials). For such low P dose the interstitial concentration is below a critical one for T>550°C so that no rod-like defects or dislocation loops can be created. Only TED related to dissolution of rod-like defects resulting in short-term TED and dislocation loop dissolution resulting in a long-term TED might be affecting our quantification method for the decay of the P concentration in the peak (δP). Error bars (Figure 4) in P-I diffusivity extraction reflects this small effect as well as other error sources. Figure 1 is a cartoon diagram showing experimental features.

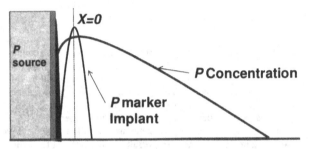

Figure 1 Diagram illustrating idea for increased observability of pair diffusion

Due to P implant and P drive-in, Si_I concentration is at the level of supersaturation while Si Vacancy (Si_v) concentration is undersaturated.

$$P_{Drive-In} \longrightarrow Si_I: \quad Supersaturation; \; Si_V - Understaturation \qquad (1)$$

During injection of I into the Si substrate, P-Si_I pairs are formed with a generation rate (τ_g)

$$P_s + Si_I \xrightarrow{\tau_g} C_{PI}: \quad Pair-Formation-(Kick-out-Mechanism) \qquad (2)$$

After P-Si_I pair formation, the pair diffuses in the Si lattice with diffusivity D_{PI}. The following differential equation governs pair formation, diffusion and dissociation.

$$\frac{\partial C_{PI}}{\partial t} = D_{C_{PI}} \nabla^2 C_{PI} + \frac{P_s}{\tau_g} - \frac{C_{PI}}{\tau_p}: \qquad (3)$$

Finally, P-Si_I pair will dissociate governed by pair lifetime (τ_p)

$$C_{PI} \xrightarrow{\tau_p} P_S + Si_I - Pair-Dissociation \qquad (4)$$

The actual experimental steps are as follows. First, P is implanted as a marker layer at a shallow depth followed by Si surface cleaning and conditioning. Next highly P doped Poly-Si is deposited by CVD at low temperature and finally P profile is measured by SIMS.

Figure 2 shows a cross sectional view highlighting interstitial injection, pair formation, diffusion and dissociation.

Interstitial Injection

```
Implanted P          Interstitial
        ⟨I⟩          Injection
     ⟨P⟩⟨P⟩⟨P⟩

Pair
Formation    ⟨P⟩—⟨I⟩

Pair Diffusion ⟨P⟩⟨I⟩

Pair
Dissociation ⟨P⟩  ⟨I⟩

Bulk Si
```

Figure 2 Cross Section of sample during experiment highlighting interstitial injection, pair formation, diffusion and dissociation

RESULTS

The phosphorus concentration after low temperature poly deposition was studied with SIMS and a typical result is plotted in Figure 3.

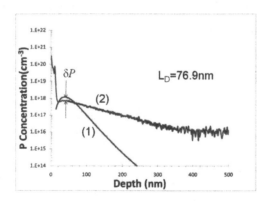

Figure 3 (1) Gaussian Profile simulated with TCAD coincides with theoretical model after P ion implant; (2) P Concentration vs Depth showing long range exponential diffusion; Curve (2) is exponentially fitted and pair diffusion length is determined

P profile shows exponential decay of two orders of magnitude for a significant depth scale. This is contrary to a Gaussian profile simulated with TCAD accounting for TED during total thermal budget of Poly-Si deposition. The results confirm that P-Si_I pairs exhibit a long life-time and that the profile is driven by long distance P-Si_I pair diffusion and dissociation into substitutional P and Si_I. Long range exponential P-Si_I pair diffusion is observable because unlike other systems [2], diffusion of phosphorous is enhanced significantly at low temperature due to interstitial injection by P drive-in thus increasing the sensitivity of the method. From this data and exponential fitting, P-Si_I diffusion length was calculated to be 76.9nm.

Model

A qualitative model based on equations (1-4), described above, can be formulated as a BVP for a system with injection of interstitials at x=0 followed by pair formation diffusion, and dissociation.

The BVP can be represented by the following continuity equation (5)

$$\frac{\partial C_{PI}}{\partial t} = D_{C_{PI}} \frac{\partial^2 C_{PI}}{\partial x^2} - \frac{C_{PI}(x) - C_{PI\,eq}}{\tau_p} \qquad (5)$$

$$C_{PI}(x=0) = C_{PI0} \qquad C_{PI}(x=\infty) = 0 \qquad (6,7)$$

Equations (6,7) are the boundary conditions for this BVP. The generation term from equation (3) above is replaced by boundary condition (6) meaning a constant rate of pair formation C_{PI} is assumed. The approximation is justified by a small decrease (δP) of peak P concentration as seen in figure 3. As a result, the P concentration at x=0 is nearly constant and the Si_I supply from P drive-in is steady except for the decaying Si_I component coming from Si_I created by P ion implantation.

In equation (5), $C_{PI,}$ D_{PI} pair and τ_p are the concentration, diffusivity and lifetime of P-Si_I pairs. The solution to the BVP is the integral,

$$C_{PI}(x,t) = A \int_0^\infty G(x'-x,t) C_{PI}(x,0) dx \qquad (8)$$

with

$$G(x',t) = \frac{C_{PI0}}{\sqrt{4\pi D_{PI} C_{PI} t}} \exp(-\frac{x'^2}{4D_{PI}t} - \frac{t}{\tau}): \qquad Green-Function \quad (9)$$

This integral can be simplified for the condition of steady state and the solution is:

$$C_{PI}(x) = (C_{PI0}) e^{(-\frac{x}{L_p})} \qquad (10)$$

The physical meaning of this approximation is beyond the scope of this paper and will be discussed elsewhere in a separate paper. Applying equation (10) to the experimental data, diffusivity and lifetime of P-Si_I pairs are calculated and presented in Table 1 with a comparison to the B system from reference [2].

	This Work for P-Si_I Pairs	Ref [2] for B-Si_I Pairs
Diffusivity of Pair	9.8×10^{-14} cm^2/sec	2.9×10^{-17} cm^2/sec
Lifetime of Pair	600 – 1800 sec	~3 x 10^4 sec
Time	~1/100x less than ref [2]	6600 min
Temperature	Similar to ref [2]	600°C

Table 1 Extraction of Diffusivity and Lifetime of pairs and compared to reference [2]

It should be noted that quasi-equilibrium P diffusivity at this low temperature cannot be extracted because super-saturation by I is unknown in this experiment. Last but not least, more research is needed to understand physical interpretation of extracted figures.

It would be useful to compare P-Si_I diffusivity with extrapolated diffusivity from high temperature range. Figure 4 plots P-Si_I pair diffusivity vs 1/T with published high temperature diffusivity is also included [1].

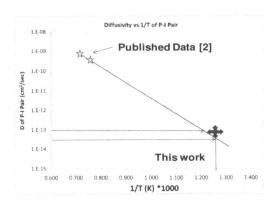

Figure 4 Diffusivity of Pair vs 1/T

Extrapolation to low temperature range shows good agreement with extracted diffusivity from this work confirming the accuracy of the P-Si_I pair diffusivity extracted. In the high temperature data there is always an uncertainty related to $D_{PI} * C_{PI}$ product. Good agreement with this data suggests that coupling of $D_{PI} * C_{PI}$ is broken down reasonably well. Please note that cross on the figure presents errors in diffusivity extraction accounting for T-t profile uncertainties and fast decaying TED as mentioned above.

Implications for SOI

The novel experiment was repeated on a SOI substrate. The results in Figure 5, confirm that the BOX-SOI interface is sink for interstitials with an extra-pile up of P at the interface due to atomic trapping. Low temperature poly-Si deposition results in an increased pile-up due to enhanced P-Si_I pair diffusion.

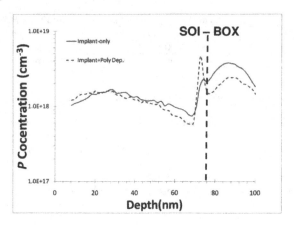

Figure 5 P Concentration before and after poly deposition on SOI Substrate showing Increased P trapping at the BOX/SOI Interface due to enhanced P-Si_I pair diffusion

CONCLUSIONS

A novel experiment has been carried out to increase observability of P-Si_I pair diffusion. A clear exponential profile is seen at low temperature, and this allowed for quantification of lifetime and diffusivity of P-Si_I pairs. This quantification allows for development of diffusion models based on physics for better TCAD predictability in modern low temperature process conditions.

REFERENCES

[1] Fahey et al. "Point Defects and Dopant Diffusion in Silicon" Reviews of Modern Physics. 61, No.2 (1989)

[2] N.E.B Cowern et al. "Impurity Diffusion via an Intermediate Species: The B-Si System" Physical Review Letters 1990 Nov 5 Vol 65 N.19 PP.2434-2437

[3] N.E.B Cowern et al. "Experiments on Atomic-Scale Mechanisms of Diffusion" Physical Review Letters 1991 July 8 Vol 67 N.2 PP.212-215

[4] V.I. Kol'dyaev, "Study of influence of the nonequlbrium point defect concentration gradient on the dopant flux during ion implantation at high temperatures" Nuclear Instruments and Methods in Phys. Res B (1995) Vol 103 PP.446-453

Mater. Res. Soc. Symp. Proc. Vol. 1155 © 2009 Materials Research Society   1155-C02-05

Role of Boron TED and Series Resistance in SiGe/Si Heterojunction pMOSFETs

Yonghyun Kim[1,2], Chang Yong Kang[2], Se-Hoon Lee[1,2], Prashant Majhi[2], Byoung-Gi Min[3], Ki-Seung Lee[3], Donghwan Ahn[1], and Sanjay K. Banerjee[1]

[1]Microelectronics Research Center, Department of Electrical and Computer Engineering, The University of Texas at Austin, Austin, Texas 78758, USA
[2]SEMATECH Inc., 2706 Montopolis Drive, Austin, Texas 78741, USA
[3]JUSUNG America Inc., 2201 Double Creek Drive Suite 5003, Round Rock, Texas, 78664, USA

ABSTRACT

We investigate boron transient enhanced diffusion (TED) and series resistance in SiGe/Si heterojunction channel pMOSFET. The stress gradient at the SiGe/Si interface near the gate edge in high Ge concentrations are found to determine boron TED as well as extension junction shape, which has a significant impact on the parasitic LDD and source/drain (S/D) series resistance. In addition, high Ge concentrations in the epitaxial SiGe layer on top of Si substrate result in a high sheet resistance during a 1000°C/5s rapid thermal processing (RTP), which is mainly due to alloy scattering and interface roughness scattering.

INTRODUCTION

Beyond the 45nm technology node, CMOS scaling has been mainly driven by high-k/metal gates and stress-induced carrier mobility enhancements [1]. To overcome intrinsic limits to carrier mobility in Si, alternative materials such as SiGe and GaAs have received the consistent attentions due to their higher carrier mobility [2-4]. Process-induced channel strain engineering is also attractive, such as compressive strain with SiGe source and drain (S/D) and channel regions in Si channel pMOSFET (PMOS) for hole mobility enhancement [5-7]. Indeed, the drive current enhancement by strain engineering and the use of high mobility materials should be maximized together, while minimizing parasitic effects such as S/D extension/overlap resistance by controlling the depth and abruptness of the extension junction in MOSFETs [8, 9].

Recently, Ranade et al. demonstrated the application of SiGe in fabricating ultrashallow junction S/D extensions in bulk Si PMOS [10]. Furthermore, King et al. showed that higher B activation can be achieved in poly-SiGe with high Ge contents compared to poly-Si for a high dose (>1e15cm^{-2}) and high energy (20keV) B implant and rapid thermal processing (RTP) [11]. Thompson et al. characterized the junction depth and sheet resistance for p$^+$ shallow junction with B-doped low temperature molecular beam epitaxy (LTMBE) and B-implanted Si/SiGe/Si layers, showing that higher Ge mole fractions up to 40% retard B diffusion significantly [12].

However, the physical mechanisms between series resistance and boron diffusion with electrical activation in the strained SiGe/Si heterojunction layer for PMOS have not been fully studied for different SiGe layer thicknesses and Ge contents (>50%), especially with high temperature annealing. While BF$_2$ implant is widely used for LDD formation in standard CMOS process, F effects [13-15] should be considered for reducing a junction depth and minimizing a sheet resistance in SiGe layers. In this study, we focus on how the device performance is affected as a function of Ge mole fraction and SiGe layer thickness in a compressively strained SiGe/Si heterojunction PMOS with a high temperature annealing.

A compressively strained SiGe/Si channel and S/D PMOS is fabricated by a standard gate-first CMOS flow. Silicon capping layer was not used for a compressively strained SiGe/Si channel and S/D PMOS. A control PMOS with a Si channel was also fabricated. Epitaxial SiGe was grown by ultrahigh vacuum chemical vapor deposition (UHVCVD) using Si_2H_6 and GeH_4 on n-type Si substrate with shallow trench isolation [16, 17]. A 20nm thickness of the epitaxial SiGe layer for PMOS was chosen based on the critical thickness to control the defects such as misfit dislocations as well as to minimize strain relaxation [18]. To characterize the sheet resistance, an epitaxial SiGe layer with 50% Ge was grown (10nm and 20nm thick) on a blanket Si wafer and BF_2 was implanted for the LDD formation condition (5 keV energy, $1e15cm^{-2}$ dose, and 7 degree tilt angle). Implanted B profiles are characterized using secondary ion mass spectroscopy (SIMS). The sheet resistance was determined by four-point probe measurement in units of Ω/square.

DISCUSSION

The transconductance for shorter channel lengths down to 60nm is shown for the control Si channel and $Si_{0.6}Ge_{0.4}$ channel in Fig. 1. The transconductance of the $Si_{0.6}Ge_{0.4}$ channel is roughly 2X better than that of Si control for a 1μm channel length in Fig. 1(a) due to the higher carrier mobility of the SiGe channel. As the channel length becomes shorter, the transconductance for both cases significantly degrades due to parasitic series resistance in PMOS. Although the transconductance of the $Si_{0.6}Ge_{0.4}$ channel is superior to that of the Si control for long channels, it is no longer valid for short channels due to higher parasitic series resistance in the LDD and S/D.

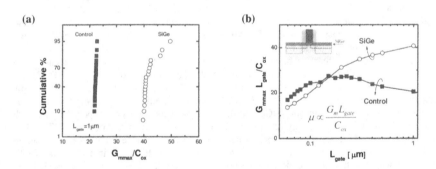

FIGURE 1. Transconductance is shown (a) for long channel (1μm) and (b) for several channel lengths down to 60nm for the Si control and SiGe/Si heterojunction PMOS. The schematic diagram of the SiGe/Si heterojunction PMOS is demonstrated in the inset. The dashed blue and dotted red lines show the possible extension and deep S/D junction, respectively.

In order to clarify the physical mechanisms, the channel conductivity and parasitic S/D resistance are calculated with 20nm SiGe thickness as a function of Ge contents in Fig. 2 [19]. The inset of Fig. 2 is shown for the contributions of the channel and parasitic resistance, respectively.

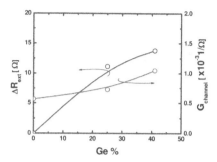

FIGURE 2. Channel conductivity and extension resistance is shown as a function of Ge contents. The schematic diagram in the inset shows channel and parasitic S/D resistance.

By increasing the amount of Ge contents, intrinsic channel conductivity becomes greater and the parasitic extension resistance is also increased. High parasitic resistance for high Ge concentrations is probably not due to boron deactivation by increased Ge concentrations because King et al. reported that SiGe could provide significantly higher active B concentrations than Si [11].

Boron diffusion in SiGe is known to be retarded during transient enhanced diffusion (TED) because Ge may play an important role in either trapping B atoms by Ge-B pairing or increasing the diffusion barrier of Si-B pair [20, 21]. In addition, the compressive strain in SiGe and the tensile strain near the Si interface from the gate edge to the channel are also a driving force for interstitial-mediated B TED because tensile stress increases the stability of interstitial defects while compressive stress decreases it [22]. Lin et al. identified that the biaxial tensile strain in Si enhances B diffusion significantly and might induce an anisotropic B diffusion in lightly p-doped conditions [23].

The strain induced in S/D and LDD are expected to be relaxed significantly with BF_2 implant and thermal annealing. However, the strain in SiGe/Si heterojunction near the gate edge is still sustained to increase or decrease B diffusion into the lateral and vertical direction. Based on the physical mechanisms explained above, B diffusion in SiGe layer is retarded due to biaxial compressive strain and Ge effect while B diffusion near Si interface slightly is increased due to biaxial tensile strain [12, 22, 23]. Therefore, SiGe/Si heterojunction in the gate overlap region has a highly distorted extension junction for high Ge concentrations, resulting in increasing the parasitic series resistance, compared to a box-shaped abrupt junction shape [8].

In order to identify the relative impact on series resistance for the extension junction shape and boron activation, the sheet resistance for the epitaxial SiGe layer on top of Si substrate is measured to investigate the effect of high temperature annealing and the dependence on Ge concentrations. In Fig. 3, sheet resistance is shown for an epitaxial SiGe layer on top of Si substrate and for a Si control. The samples with 50% Ge concentrations have a higher sheet resistance than the Si control with a 1000°C/5s RTP. For a 1000°C/30s RTP, the sheet resistance becomes to be reduced, but it is still high for high Ge concentrations.

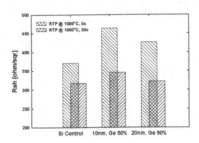

FIGURE 3. Sheet resistance for 10nm and 20nm $Si_{50}Ge_{50}$/Si heterojunction and for the Si control. RTP is done for 5s and 30s at 1000°C after BF_2 implant.

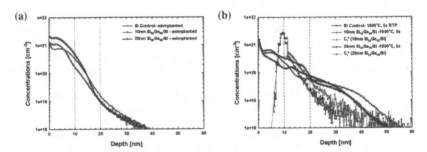

FIGURE 4. SIMS profiles (a) for as-implanted samples and (b) for 1000°C/5s RTP samples. Initial interstitial concentrations (C_I^*) after the BF_2 implant are approximated using Kinetic Monte Carlo (KMC) process simulation.

Recent studies showed that sheet resistance is increased due to F counter doping effect when B profile is overlapped with F profile in Si [13, 15]. While the binding energy between F-Ge is assumed to be low compared to that of F-Si, F counter doping effect is also expected to exist in SiGe layers with BF_2 implant. In addition, Xia et al. reported that enhanced Si-Ge interdiffusion increases alloy scattering when the compressive strain in the SiGe layer is induced in the Si/SiGe/Si heterojunction [24]. High sheet resistance in the $Si_{50}Ge_{50}$/Si heterojunction could be originated from alloy scattering and from interface roughness scattering due to Ge outdiffusion. The results in Fig. 2 show that high Ge concentrations increase a sheet resistance in SiGe/Si heterojunction and, in turn, increase parasitic series resistance. Therefore, high Ge concentrations in the epitaxial SiGe layer on top of Si are not always desirable because of the high sheet resistance.

SIMS measurements were performed on as-implanted and annealed samples for 10nm and 20nm $Si_{50}Ge_{50}$ layers on top of Si substrate (Fig. 4). A kinetic Monte Carlo (KMC) process simulation [25] was also performed. The dotted vertical line shows the depth of the SiGe layer.

Even though the initial profiles are almost identical for all the samples shown in Fig. 4(a), B diffusion was retarded at the interface of $Si_{50}Ge_{50}$ and Si (Fig. 4(b)). The initial interstitial defect profiles (C_I^*) after the BF_2 implant were approximated by KMC process simulation. The position of maximum C_I^* indicates the approximate amorphous and crystalline (a/c) interface near which maximum damages in $Si_{50}Ge_{50}$/Si heterojunction are generated. Since C_I^* is located near the interface of 10nm SiGe layer, the interface roughness scattering of 10nm SiGe layer is increased compared to that of 20nm SiGe layer and then the sheet resistance becomes to be higher for the 10nm SiGe layer than the 20nm SiGe layer (Fig. 3) [24, 26].

CONCLUSIONS

We characterized series resistance in a compressively strained SiGe/Si heterojunction PMOS. The performance degradation for SiGe/Si heterojunction PMOS with shorter channel lengths is mainly caused by a parasitic LDD and S/D series resistance, especially under higher Ge concentrations. The parasitic resistance is mainly originated from due to both the highly curved junction shape and interface roughness scattering in SiGe/Si heterojunction. In order to have a box shaped LDD junction and reduce a scattering-induced degradation, SiGe thickness and Ge concentrations are required to be optimized. Because F counter doping effect might increase a sheet resistance, it is suggested that F doping should be utilized to control the point defects and B TED with avoiding the profile overlap between B and F in SiGe/Si heterojunction. In the conventional CMOS process, since RTP is commonly used for implant damage annealing and dopant activation, the parasitic LDD and S/D series resistance is expected to become an important contributor to device degradation. Thus, the physical understanding afforded by our study is helpful in minimizing the parasitic series resistance in SiGe/Si heterojunction PMOS.

ACKNOWLEDGMENTS

This work is supported by the Semiconductor Research Corporation (SRC). Y. Kim would like to thank Applied Materials Graduate Fellowship program for financial support. This work was also partially supported by the "System IC 2010" project of the Korea Ministry of Knowledge Economy.

REFERENCES

[1] T. Ghani, M. Amstrong, C. Auth, M. Bost, P. Chavat, G. Glass, T. Hoffman, K. Johnson, C. Kenyon, J. Klaus, B. McIntyre, K. Mistry, A. Murthy, J. Sandford, M. Silberstein, S. Sivakumar, P. Smith, K. Zawadzki, S. Thompson, and M. Bohr, in IEDM Tech. Dig., 1161 (2003).
[2] A. G. O'Neill and A. D. Antoniadis, IEEE Trans. Electron Devices 43, 911 (1996).
[3] K. Rim, J. L. Hoyt, and J. F. Gibbons, IEEE Trans. Electron Devices 47, 1406 (2000).
[4] S. Lee, P. Majhi, J. Oh, B. Sassman, C. Young, A. Bowoner, W. Loh, K. Choi, B. Cho, H. Lee, P. Kirsch. H. Harris, W. Tsai, S. Datta, H. Tseng, S. K. Banerjee, and R. Jammy, IEEE Electron Device Lett. 29, 1017 (2008).
[5] R. People, IEEE J. Quantum Electron 22, 1696 (1986).
[6] J. J. Welser, J. L. Hoyt, and J. F. Gibbons, IEEE Electron Device Lett. 15, 100 (1994).
[7] G. H. Wang, E. Toh, A. Du, G. Lo, G. Samudra, and Y. Yeo, IEEE Trans. Electron Devices 29, 77 (2008).
[8] S. Kim, C. Park, and J. C. S. Woo, IEEE Trans. Electron Devices 49, 467 (2002).
[9] S. Kim, C. Park, and J. C. S. Woo, IEEE Trans. Electron Devices 49, 1748 (2002).

[10] P. Ranade, H. Takeuchi, W. Lee, V. Subramanian, and T. King, IEEE Trans. Electron Devices **49**, 1436 (2002).

[11] T. J. King, J. McVittie, K. C. Saraswat, and J. R. Pfiester, IEEE Trans. Electron Devices **41**, 228 (1994).

[12] P. E. Thompson, R. Crosby, J. Bennet, and S. Felch, J. Vac. Sci. Technol. B **22(5)**, 2333 (2004).

[13] J. Park, Y. Huh, and H. Hwang, Appl. Phys. Lett. **74**, 1248 (1999).

[14] N. E. B. Cowern, B. Colombeau, J. Benson, A. J. Smith, W. Lerch, S. Paul, T. Graf, F. Cristiano, X. Hebras, and D. Bolze, Appl. Phys. Lett. **86**, 101905 (2005).

[15] G. Impellizzeri, S. Mirabella, A. M. Piro, M. G. Grimaldi, F. Priolo, F. Giannazzo, V. Raineri, E. Napolitani, and A. Carnera, Appl. Phys. Lett. **91**, 132101 (2007).

[16] B. Anthony, T. Hsu, L. Breaux, R. Qian, S. K. Banerjee, and A. Tasch, J. Elec. Mat. **19**, 1027 (1990).

[17] C. Li, S. John, E. Quinones, and S. K. Banerjee, J. Vac. Sci. Tech. A **14**, 170 (1996).

[18] R. People and J. C. Bean, Appl. Phys. Lett. **47**, 322 (1985).

[19] Yuan, T. and H.N. Tak, Fundamentals of modern VLSI devices. 1998: Cambridge University Press. 469.

[20] P. Kuo, J. L. Hoyt, J. F. Gibbons, J. E. Turner, and D. Lefforge, Appl. Phys. Lett. **66**, 580 (1995).

[21] N. R. Zangenberg, J. Fage-Pedersen, and J. L. Hansen, J. Appl. Phys. **94**, 3883 (2003).

[22] S. Dunham, M. Diebel, C. Ahn, and C. Shih, J. Vac. Sci. & Tech. **24**, 456 (2006).

[23] L. Lin, T. Kirichenko, B. R. Sahu, G. S. Hwang, and S. K. Banerjee, Phy. Rev. B **72**, 205206 (2005).

[24] G. Xia, O. Olubuyide, J. L. Hoyt, and M. Canonico, Appl. Phys. Lett. **88**, 013507 (2006).

[25] Sentaurus Process User Guide A-2008.09, Mountain View, Synopsys, 2008.

[26] R. T. Crosby, K. S. Jones, M. E. Law, L. Radic, P. E. Thompson, and J. Liu, Appl. Phys. Lett. **87**, 192111 (2005).

Mater. Res. Soc. Symp. Proc. Vol. 1155 © 2009 Materials Research Society 1155-C01-01

Epitaxial Lanthanide Oxide Based Gate Dielectrics

H. Jörg Osten[1], Apurba Laha[1], and Andreas Fissel[2]

[1] Institute of Electronic Materials and Devices, Leibniz University Hannover, Appelstr. 11A
[2] Information Technology Laboratory, Leibniz University Hannover, Schneiderberg 32
D-30167 Hannover, Germany

ABSTRACT

Many materials systems are currently under consideration as potential replacements for SiO_2 as the gate dielectric material for sub-0.1 μm CMOS technology. We present results for crystalline gadolinium oxides on silicon in the cubic *bixbyite* structure grown by solid source molecular beam epitaxy. On Si(100), crystalline Gd_2O_3 grows usually as (110)-oriented domains, with two orthogonal in-plane orientations. Layers grown under best vacuum conditions often exhibit poor dielectric properties due to the formation of crystalline interfacial silicide inclusions. Additional oxygen supply during growth improves the dielectric properties significantly. Layers grown by an optimized MBE process display a sufficiently high-K value to achieve equivalent oxide thickness values < 1 nm, combined with ultra-low leakage current densities, good reliability, and high electrical breakdown voltage. A variety of MOS capacitors and field effect transistors has been fabricated based on these layers. Efficient manipulation of Si(100) 4° miscut substrate surfaces can lead to single domain epitaxial Gd_2O_3 layer. Such epi-Gd_2O_3 layers exhibited significant lower leakage currents compared to the commonly obtained epitaxial layers with two orthogonal domains. For capacitance equivalent thicknesses below 1 nm, this differences disappear, indicating that for ultrathin layers direct tunneling becomes dominating. We investigated the effect of post-growth annealings on layer properties. We showed that a standard forming gas anneal can eliminate flatband instabilities and hysteresis as well as reduce leakage currents by saturating dangling bond caused by the bonding mismatch. In addition, we investigated the impact of rapid thermal anneals on structural and electrical properties of crystalline Gd_2O_3 layers grown on Si with different orientations. The degradation of layers can be significantly reduced by sealing the layer with amorphous silicon prior to annealing.

MATERIAL SELECTION

Many materials systems are currently under consideration as potential replacements for SiO_2 as the gate dielectric material for sub-0.1 μm CMOS technology. The upcoming generation of high-K dielectrics will be most likely formed by amorphous hafnium-based alloys. However, the existence of an interfacial layer of SiO_2 or another low permittivity material, will limit the highest possible gate stack capacitance, or equivalently, the lowest achievable equivalent oxide thickness (*EOT*) value. Thus, any interfacial lower-permittivity layer should be minimized in future generations. It is essential to keep channel carrier mobilities high. In addition, the increased process complexity for the deposition and control of additional ultrathin dielectric layers, as well as scalability to later technology nodes, remains a concern. The 2nd generation of high-K materials have to be formed without any interfacial layer to realize $EOT < 0.8$ nm. In such a case, the high-K dielectric/Si interface properties influence the device performance significantly. A good interface requires either that the oxide is amorphous, or that it is epitaxial and lattice-matched to the underlying silicon. Amorphous dielectrics are expected to be able to adjust the local bonding to minimize the number of Si dangling bonds at the interface. The alternative is to use an epitaxial oxide. This involves more effort, but it has the advantage of enabling defined interfaces engineering. Generally, there are two groups of possible candidates for epitaxial growth on Si, namely (a) perovskite-type structures and (b) binary metal oxides, in particular lanthanide oxides [1,2]. In the following, we will concentrate on binary lanthanide oxides.

Lanthanide oxides (LnO's) form a very interesting group of insulators for epitaxial growth on silicon [3]. It is extremely important and desirable to integrate these highly functional metal oxides into mature semiconductor technologies. The LnO's can have different oxygen compositions LnO_x, with x ranging from 1 to 2 due to the multiple oxidation states (+2, +3, and +4) of the rare-earth metals [4]. This leads to oxides with different stoichiometries (LnO, Ln_2O_3, LnO_2). All known Ln (II) oxides, like EuO, are not insulating. Therefore, they will not be considered here any further. For application in a Si-based device fabrication process, all lanthanide oxides exhibiting more than one valence state (+3 and +4) are not the best choice as epitaxial high-K materials because of the coexistence of phases with different oxygen content. For example, cerium(IV) oxide (CeO_2) can release oxygen under reduction conditions forming a series of reduced oxides with stoichiometric cerium(III) oxide (Ce_2O_3) as an end product, which in its turn easily takes up oxygen under oxidizing conditions, turning the cerium(III) oxide back into CeO_2. In addition, stable mixed valence-state structures can occur for some LnO's. For example, the mixed valence-state Pr_6O_{11} is the most stable phase for praseodymium oxide. For those highly ionic oxides, the position of the charge neutrality level depends strongly on the stoichiometry [5]. Thus, also the band alignment to silicon and, finally, the leakage behavior becomes strongly dependent on the oxygen content. All lanthanide oxides displaying only one valence state are easier to handle due to the absence of transitions between phases with different oxygen content. Based on that argument, we will focus our discussion mainly on lanthanide (III) oxides (occurring as Ln_2O_3).

The Ln_2O_3 oxides can occur in different structural phases, like the manganese oxide (Mn_2O_3) or *bixbyite* structure. Some oxides also crystallize in the hexagonal lanthanum oxide structure, which is suitable for epitaxy only on Si(111). Also, monoclinic phases are known for various lanthanide (III) oxides. Different crystallographic structures are accompanied by different dielectric properties. Several lanthanide oxides can undergo structural phase transformation within a temperature range, typical for CMOS processing [6]. Also these oxides are not very well suited for technological applications. Epitaxial growth on a clean surface requires matching in symmetry as well as in atomic spacing. Here, we will present results for crystalline gadolinium oxide on silicon with the Gd_2O_3 composition in the cubic *bixbyite* structure which has a large band gap of about 6 eV and nearly symmetrical band offsets to Si [7] as well as a low lattice mismatch of about 0.5 %. The *bixbyite* structure is based on the calcium fluorite structure, where 1/4 of the oxygen atoms have been removed from specific lattice sites. That structure has a lattice symmetry suitable for epitaxial growth on Si(100) and Si(111).

EPITAXY OF LANTHANIDE OXIDE ON SILICON(100)

Commonly, epitaxial heterostructures are evaluated on the basis of lattice matching, with the misfit defined as the relative difference in the lattice constants (a_{film} and a_{Si}). Here, the lattice mismatch would be identical for all three epitaxial relationships, i.e. (100)//(100), (110)//(110), and (111)//(111). All other combinations violate symmetry matching. This concept is misleading, because epitaxial growth is also governed by the surface and interface energetics. In case of rare-earth oxides, the surface energy of the (100) surface is much higher than that of Si(100) [8]. That could result in a 3-dimensional growth mode or in a change of the growth direction towards a low-energy surface orientation. The later is observed for the growth on Si(100) for most of the lanthanide oxides, where the oxides were found to grow in (110) orientation. For the understanding of the Ln_2O_3(110)/Si(100) interface formation we have to consider that the lattice is made up of metal atoms occupying the positions of a face-centered cubic lattice with a lattice constant a_{film}, where the tetrahedral holes are occupied by oxygen atoms. However, the growth of LnO's is based on the deposition of metal oxide molecules. Due to the existence of highly ionic Ln—O bonds in combination with the high bonding strength of the covalent Si—O bonds, we can assume that the interface is predominantly formed by Si—O—Ln bonds. Therefore, the matching of the oxygen atoms is the important parameter. The complete crystallographic structure can also be described by two non-identical metal and oxygen lattices, respectively. The arrangement of the oxygen atoms forms a simple cubic lattice with a lattice constant of $a_{film}/2$. The interesting matching condition for epitaxial growth is the Ln_2O_3(110)[100]//Si(100)[110] relation. In that case, nearly 1:1 matching should occur along one direction. In the other direction, there would be roughly a 3:2 matching relation. Layers grown in this orientation exhibit mostly two types of (110)-oriented domains, with two orthogonal in-plane orientations (found experimentally for a large variety of binary metal oxides on Si(100), like Er_2O_3 [9], Sm_2O_3 [10], Lu_2O_3 [11,12], Sc_2O_3 [13], Gd_2O_3 and Y_2O_3 [14]). That is even enforced by the Si(100) dimer (2x1) surface reconstruction. Dimer rows on adjacent terraces are oriented perpendicular to each other and nucleation of the oxide on Si(100) follows the dimer orientation, what results in the domains on stepped surfaces [15]. Due to the 45° rotation of the (110) plane relative to the substrate, the mismatch for the 1:1 matching is identical to that one obtained from applying the lattice constant concept.

GROWTH OF CRYSTALLINE GADOLINIUM OXIDES ON SILICON

All experiments were performed in a multi-chamber ultrahigh-vacuum Molecular Beam Epitaxy (MBE) system (*DCA Instruments*) capable to handle 8" wafer. This system includes a growth, an annealing, and an analysis chamber connected by an ultra-high vacuum (UHV) transfer system. The layers were grown on 4" Si substrates with different orientations. Substrates were cleaned *ex situ* using as the last step diluted HF etch (HF:H_2O = 1:10) followed by a dilution rinse, and then were immediately inserted into the vacuum system. Substrates were annealed *in situ* to transform the initial hydrogen-terminated (1×1) surface structure into the (2×1) superstructure indicating a clean and well-ordered surface. Commercially available, granular Gd_2O_3 material was evaporated using an electron-beam evaporator. Growth temperatures were in the range 800-1000 K. Typical growth rates were 0.005 – 0.01 nm/sec. The surface and layer structure was evaluated during the growth by reflection of high energy electron diffraction (RHEED). After growth, the wafers can be transferred into the analysis chamber without leaving the UHV to perform x-ray photoelectron spectroscopy (XPS) investigations.

The layer thickness was measured *ex vacuo* by X-ray reflectivity (XRR) using a standard single crystal diffractometer with graphite monochromator in front of the detector. Layers were also characterized by X-ray diffraction (XRD) (Θ-2Θ, ω and Φ-scans) and transmission electron microscopy (TEM), i.e., high resolution cross-section and plan-view images combined with selected area diffraction (SAD).

On Si(100) substrate, the grown layers exhibit the known two types of (110)-oriented domains, with two orthogonal in-plane orientations [16]. On (111) oriented Si surface, *bixbyite* Gd_2O_3 grows epitaxially when the substrate temperature is above 600 °C. Here, the oxide exhibits an A/B twining relationship where B is related to the substrate twinning orientation A by a 180° rotation around the Si(111) surface normal. On a Si(110) surface, grown layers exhibit pronounced faceting by the development of low-energy {111} facets [17]. Figure 1 shows the X-ray (Θ-2Θ scan) diffraction patterns of Gd_2O_3 thin films grown on different Si substrates. Gd_2O_3 layer on Si(100) surface exhibits a distinct peak at 2θ=46.4° corresponding to the d(440) inter planar spacing of cubic Gd_2O_3 along [110] orientation. The layers on the Si(111) substrate also exhibit single orientation without any indication of disoriented crystallites. The peaks at 28.5° and 59.4° correspond to d(222) and d(444) interplanar spacing are hidden under the appropriate Si peaks. The Θ-2Θ scan for Gd_2O_3 thin films grown on Si(110) displays a peak at 28.5° confirming the preferential growth along (111) direction as observed in RHEED images (not shown).

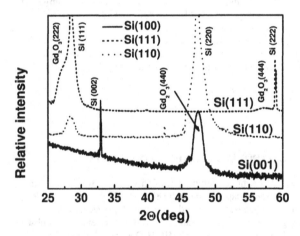

Figure 1: X-ray diffraction patterns (Θ-2Θ scan) of Gd_2O_3 thin films grown on silicon substrates with different orientations

IMPACT OF OXYGEN CONCENTRATION ON LAYER PROPERTIES

Recently, we investigated the interface and layer formation processes of Ln_2O_3 films on Si(100) grown under ultra-clean ultra-high vacuum (UHV) conditions of MBE [18]. Layers grown under best vacuum conditions often exhibit very high leakage currents. We found that the partial oxygen pressure during the interface formation and during growth is a very crucial parameter. Too low oxygen content can lead to the formation of silicide-like inclusions. For Ln_2O_3 growth, the formation and stability of the silicide-like phase depends on the oxygen chemical potential [19]. Considering the low oxygen partial pressure under UHV conditions (as used in MBE growth), the chemical potential of oxygen can become negative. Thus, silicide formation will be favorable compared to oxide formation. That is one of the most crucial points for the growth of dielectric layers, because the silicide growth can continue as long as the oxygen content remains low enough or the oxygen chemical potential remains strongly negative. This occurs even faster at a surface or interface region where the energetical equilibrium is distorted, for example, due to stress. The appearance of silicide inclusions seems to be a serious drawback for a lot of possible material candidates for future high-K applications. For the epitaxial growth of Ln_2O_3 oxides, a well controlled interface and subsequent growth engineering is necessary to reach the required electrical properties. The available oxygen concentration can be controlled by modifying MBE growth processes using an additional oxygen supply. We found, that MBE growth under defined oxygen partial pressures during the interface formation and/or during the subsequent growth can prevent any kind of silicide inclusions and void formation [20]. On the other hand, too high oxygen content might oxidize the Si surface, leading to a lower-K interfacial SiO_x layer, and finally limits the achievable minimal equivalent oxide thickness.

In a set of experiments, we varied the growth temperature while keeping all other parameters constant. Figure 2 shows the extracted CET values versus the physical oxide thickness. The oxygen partial pressure during growth was kept at $1 \cdot 10^{-7}$ mbar. If a structure contains several dielectrics in series, the lowest capacitance layer will dominate the overall capacitance, and also set a limit on the minimum achievable EOT value. For example, the total capacitance of two dielectrics in series is given by

$$1/C_{tot} = 1/C_1 + 1/C_2, \qquad (1)$$

where C_1 and C_2 are the capacitances of the two layers, respectively. If one considers a dielectric stack structure such that the bottom layer (layer 1) of the stack is SiO_2, and the top layer (layer 2) is the high-K alternative gate dielectric, Eq. (1) is simplified (assuming equal areas) to

$$t_{eq} = t_{SiO2} + (K_{ox} / K_{high-K}) t_{high-K}. \qquad (2)$$

Electrical measurements of such a multilayer stack will yield always an effective dielectric constant. In a simple picture, a system including an interfacial layer and a high-K layer can be described by a linear relation between CET and the physical thickness (Eq. (2)).

The experimental results shown in Fig. 2 follow that relation. The slope for all three temperatures is nearly equal resulting in an intrinsic dielectric constant of around 20 for the high-K layer in all three cases. Only the intercept for t = 0 varies with growth temperature, yielding 0.4 nm, 0.9 nm, and 1.4 nm, respectively. Based on Eq. (2), this intercept is often attributed to the physical thickness of an interfacial layer only. However, an increase of the thickness with increasing growth temperature was not supported by XRR and XTEM investigations. Instead, we suppose that the permittivity (K_{IF}) at the interface decreases due to the transformation of the interfacial transition from a silicate-like type to a more silicon-oxide like. That can be explained by the fact that gadolinium, in contrast to cerium or praseodymium, can only exist in the +3 oxidation state. Thus under equilibrium conditions, the Gd_2O_3 bulk cannot act as an effective source for oxygen supply to the Si/dielectric interface. For interfacial layer being not silicon dioxide, Eq.(2) transforms into

$$CET = t_{IF} (3.9/K_{IF}) + (3.9/K_{high\text{-}K}) \, t_{high\text{-}K}. \tag{3}$$

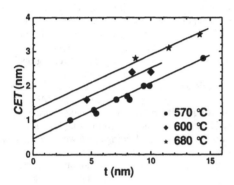

Figure 2: *CET* as a function of the physical layer thickness for Gd_2O_3 layers grown on p- and n-type Si(100) at 570 °C, 600 °C, and 680 °C, respectively. The oxygen partial pressure during growth was kept at $1 \cdot 10^{-7}$ mbar.

Next, we investigated the influence of different oxygen partial pressures. The results are very similar for all investigated growth temperatures. For the layer grown with the lowest oxygen pressure ($1 \cdot 10^{-7}$ mbar), we always found strong hysteresis which we attribute to the presence of oxygen vacancies. Increasing the oxygen pressure to $5 \cdot 10^{-7}$ mbar results in a significant reduction of the hysteresis. Further increase in oxygen partial pressure does not lead to a further improvement of electrical layer properties, but increases the physical thickness of the interfacial layer, detected by XTEM. At this pressure, oxygen concentration becomes supersaturated at the growth front, and oxygen atoms diffuse to the interface an starts to form an silicon oxide-like interfacial layer.

We obtained the best electrical results for growth at 600 °C and $p_{O2} = 5 \cdot 10^{-7}$ mbar. Figure 3 shows a cross-sectional TEM micrograph of a sample with a 15 nm thick Gd_2O_3 layer grown under these conditions on Si(100). No pronounced interfacial layer between the Si and the oxide layer can be detected.

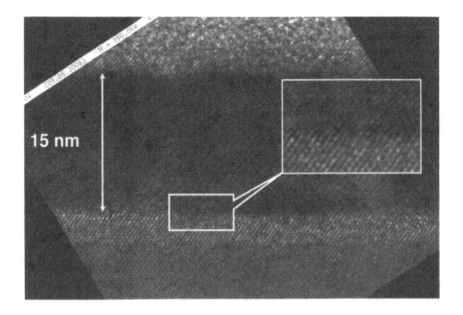

Figure 3: High-resolution cross-sectional TEM image of a Gd_2O_3/Si(100) stack. The oxide layer was grown at 600 °C and $p_{O2} = 5 \cdot 10^{-7}$ mbar. No pronounced interfacial layer is visible.

HIGH-*K* APPLICATIONS

Layers grown by an optimized MBE process display a sufficiently high-*K* value to achieve equivalent oxide thickness values < 1 nm, combined with ultra-low leakage current densities, good reliability, and high electrical breakdown voltage (Fig. 4) [20,21]. This makes epitaxial Gd_2O_3 layers excellent alternative for replacing SiO_2 as a gate dielectric [22]. First electrical characteristics of fully-depleted n- and p-type SOI-MOSFETs with epitaxial Gd_2O_3 and TiN metal gate electrodes demonstrate the feasibility of this novel gate insulator [23]. Electrical properties of the gate stack have been extracted from the devices and CMOS process compatibility has been addressed. The oxide interface state density has been found to decrease after rapid thermal annealing. On the other hand, mobile oxide charges and oxide traps have been significantly reduced by RTA. The p-MOSFETs exhibit higher saturation currents compared to n-MOSFETs as they were accumulation-mode transistors with a p^+–p^-–p^+ doping structure.

In accumulation-mode p-MOSFETs, higher drain currents can be achieved as conduction appears at the front interface as well as in the body region below compared to an inversion channel which is confined at the front interface [23]. In order to minimize process induced oxide damage, MOSFET's with W/Gd_2O_3 gate stack have been fabricated in a virtually damage-free damascene metal process [24]. The major process steps of the replacement gate process are the following: Initially, dummy gate stacks are formed followed by self-aligned S/D ion implantation. An alignment-oxide is deposited by chemical vapor deposition (CVD), annealed and planarized by chemical-mechanical polishing (CMP) down to the gate level. The dummy gates are removed completely by wet chemical etching, leaving a self-aligned imprint of the gate stack on the Si-wafer. Subsequently, the gate dielectric (Gd_2O_3) is grown. A metal layer is deposited on top of the gate dielectric and CMP is used to pattern the damascene metal gates. Standard back-end processing completes the fabrication. This was the first successful attempt to integrate crystalline high-K dielectrics into a "gentle" damascene metal gate process in order to reduce process induced oxide damages.

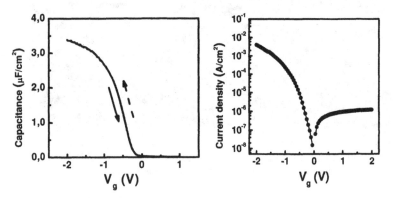

Figure 4: MOS capacitor with Gd_2O_3 on p-Si(100): EOT < 0.7 nm, hysteresis < 10 mV , J @ (V_{FB}-1V) = 0.5 mA/cm^2,

Although the lattice mismatch between Gd_2O_3 and Si is small, we have to face a binding mismatch at the Gd_2O_3/Si(100) interface, i.e. a large number of dangling bonds. The as-grown single crystalline Gd_2O_3 thin films often suffers from flatband voltage instability and large hysteresis which are possibly due to the these intrinsic dangling bonds. The instability of flatband voltage and hysteresis of Pt/Gd_2O_3/Si and W/Gd_2O_3/Si structures can be fully eliminated by the introduction of traditional forming gas annealing with proper process optimization [25]. Both optimized metal-oxide-semiconductor structures show negligible hysteresis with the interface state at the magnitude order of 10^{11} /cm^2 eV at the midgap of silicon and can be considered for the future of complementary metal oxide semiconductor devices.

In addition, we investigated the impact of rapid thermal anneals on structural and electrical properties of crystalline Gd_2O_3 layers grown on Si with different orientations [26]. The degradation of layers can be significantly reduced by sealing the layer with a-Si prior to annealing. For the capped layers, the effective capacitance equivalent thickness increases only slightly even after a 1000 °C anneal.

EFFECT OF DOMAIN BOUNDARIES

Despite having crystalline structure, such epitaxial rare earth oxides, when grown on Si(100) substrates, usually exhibit two (110) oriented orthogonal domains and hence, one could anticipate a great impact of domain boundaries on final electrical performance. However, to elucidate the impact of such domain boundaries is more complicated because the growth of epitaxial lanthanide oxides in the common *bixbyite* structure on standard Si(100) substrates always leads to the formation of these two types of [110]-oriented domains, with two orthogonal in-plane orientation, each of them exhibiting two fold symmetry. It was suggested by Kwo *et al.* [14,27] that the use of vicinal (4° miscut along <110> azimuth) Si(100) substrate could be a viable way to grow single-domain (SD) epitaxial layer, thus to eliminate domain boundary effect on electrical properties similar to that was also known from the growth of III–V compounds on Si(100) [28].

Recently we showed that not only the use of vicinal surface is necessary but also the preparation of silicon surface prior to the layer growth is a very important step to achieve single domain epitaxial Ln_2O_3 layer on vicinal (4° miscut) Si(100) substrates [29]. The reason could be understood in the following way. It was reported earlier that the presence of dimers on a (2x1) reconstructed Si(100) surface forces the overgrown lanthanide oxide layer to form orthogonal oriented ad-dimers in line with Si dimer rows [30]. Normally, dimers rows on adjacent terraces are oriented perpendicular to each other and nucleation of the oxide on (2x1) Si(100) follows the dimmer orientation, what results in the domains. If, this makes an arguable explanation; then, it seems to be possible to grow a single domain (SD) epi-layer on Si substrate if and only if all dimers on the surface are parallel to each other. Obviously, the next question is how to achieve such Si(100) surfaces with only one dimers orientation? Double-atomic steps results the Si-dimers arranged only in one orientation across the whole surface. Such steps are easier to achieve on 4°-miscut surface; however, it demands a careful preparation of the Si(100) surface (details can be found in Ref. 29).

Various layers were grown under identical deposition conditions (substrate temperature of 675°C with oxygen partial pressure of $5*10^{-7}$ mbar) on Si(100) 4° off substrates with and without surface preparation. The detail structural quality of the layers was investigated by x-ray diffraction technique. The asymmetric 360° x-ray phi (Φ) scans (azimuthal rotation) were carried out to determine the in-plane symmetries of the Gd_2O_3 layers and therefore, to confirm the presence of only a single or/and double domains in the layers. Figure 5 (a) and (b) compare the 360° Φ scan along the surface normal for the in-plane component of the {222} reflection of the (110) single and double domain epi-Gd_2O_3 layers of thickness 15nm and 14nm, respectively. The scan shown in Fig. 5(a) clearly exhibits twofold in-plane symmetry, corresponding to a SD structure, while Fig. 5(b) displays four peaks, separated by 90°. The intensity of the peaks from two-domain structure differs, whereas peaks from SD structure, separated by 180° have nearly the same intensity. From the these investigations (and additional TEM investigations, not shown here), we can confirm that domain boundaries could be eliminated completely in the epitaxial single crystalline lanthanide oxide grown on carefully prepared 4°-off oriented Si(100) substrates.

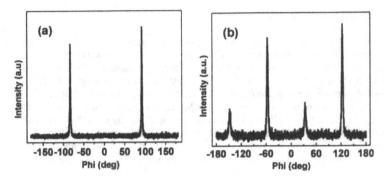

Figure 5: 360° Φ scan along the surface normal for the in-plane component of the {222} reflection of the (110) (a) SD and (b) DD epi-Gd_2O_3 layers of thickness 15 and 14 nm, respectively.

Next we performed electrical evaluations on those layers. Figure 6 compares the leakage current density of as-grown Gd_2O_3 SD and DD layers as a function of capacitance equivalent oxide thickness (CET). It clearly demonstrates that the as-grown SD Gd_2O_3 layers exhibit much lower leakage current for similar CET values, inferring that the domain boundaries in DD Gd_2O_3 layers act as the leakage paths for charge carriers. However, this disparity in the leakage current could only be observed for the thicker layers (>6 nm). For capacitance equivalent thickness below 1 nm, such differences disappear, indicating that for ultra thin layers, direct tunneling becomes dominating.

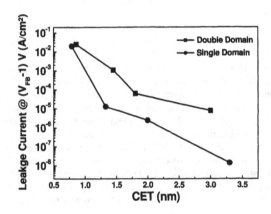

Figure 6: Leakage current density of as-grown Gd_2O_3 SD and DD layers as a function of capacitance equivalent oxide thickness (*CET*)

SUMMARY AND OUTLOOK

We described the use of molecular beam epitaxy to grow epitaxially lanthanide oxide thin films on silicon. We presented results for crystalline gadolinium oxide in the cubic *bixbyite* structure. On Si(100) oriented surfaces, crystalline Gd_2O_3 grows as (110)-oriented domains, with two orthogonal in-plane orientations. We obtain perfect epitaxial growth of cubic Gd_2O_3 on Si(111) substrates. Layers grown under best vacuum conditions often exhibit very high leakage currents. Deliberated oxygen supply during growth improves the dielectric properties significantly; however too high oxygen partial pressures lead to an increase in the lower permittivity interfacial layer thickness, and finally limit the achievable minimal equivalent oxide thickness. Experimental results for Gd_2O_3-based MOS capacitors grown under optimized conditions show that these layers are excellent candidates for application as very thin high-K materials replacing SiO_2 in future MOS devices. The minimum capacitance equivalent thickness estimated for $Pt/Gd_2O_3/Si$ MOS structures was below 0.7 nm with leakage current density of only few mA/cm^2 at $(V_g-V_{FB}) = 1$ V. We investigated the effect of post-growth annealings on layer properties. We showed that a standard forming gas anneal can eliminate flatband instabilities and hysteresis as well as reduce leakage currents by saturating dangling bond caused by the bonding mismatch. In addition, we investigated the impact of rapid thermal anneals on structural and electrical properties of crystalline Gd_2O_3 layers grown on Si with different orientations. The degradation of layers can be significantly reduced by sealing the layer with a-Si prior to annealing. For the capped layers, the effective capacitance equivalent thickness increases only slightly even after a 1000 °C anneal.

Efficient manipulation of Si(100) 4° miscut substrate surfaces can lead to single domain epitaxial Gd_2O_3 layer. Such epi-Gd_2O_3 layers exhibited significant lower leakage currents compared to the commonly obtained epitaxial layers with two orthogonal domains. For capacitance equivalent thicknesses below 1 nm, this differences disappear, indicating that for ultrathin layers direct tunneling becomes dominating.

The ability to integrate crystalline dielectric barrier layers into silicon structures can open the way for a variety of novel applications (see for example Ref. 31): Double-barrier structures comprising epitaxial insulator as barriers and Si as quantum-well are interesting because of its applicability for tunneling devices. Furthermore, quantum confinement of charged particles within nanostructured materials could lead to a large number of new phenomena, which could never been realized in normal bulk materials. Such effect could pave the way for new generation of solar cells with ultrahigh efficiencies. On the other hand, Si nano clusters embedded into insulating layers could be one of the potential contenders for nonvolatile memory device applications.

Acknowledgements
This paper summarizes part of the work we have been doing over the last years. We are in particular grateful to E. Bugiel, M. Czernohorsky, R. Dargis, J. Krügener, D. Schwendt, D. Tetzlaff, and J.X. Wang for their various contributions. We are also grateful to our partners all over the world for their support and collaboration. Part of this work was supported by the German Federal Ministry of Education and Research (BMBF) under the KrisMOS and the MegaEpos projects.

REFERENCES

1. D.P. Norton, Mat. Sci & Engineer. R **43**, 139 (2004).
2. G.-Y. Adachi and N. Imanaka, Chem. Rev. **98**, 1479 (1998).
3. H.J. Osten, M. Czernohorsky, R. Dargis, A. Laha, D. Kühne, E. Bugiel, and A. Fissel, Microelectronic Engineering **84**, 2222 (2007).
4. *The Oxide Handbook,* 2nd Edition, ed. G.V. Samsonov (IFI/Plenum, New York 1982); G.Y. Adachi and N. Imanaka, Chem. Rev. **98**, 1479 (1998).
5. J. Robertson and K. Xiong, Topics in Appl. Phys., **106**, 313 (2007).
6. M. Foëx and J.P. Traverse, Rev. Int. Hautes Temp. Refract. **3**, 429 (1966).
7. M. Badylevich, S. Shamuilia, V. V. Afanas'ev, A. Stesmans, A. Laha, H. J. Osten, and A. Fissel, Appl. Phys. Lett. **90**, 252101 (2007).
8. M. Nolan, S. Grigoleit, D.C. Sayle, St.C. Parker G.W. Watson, Surf. Sci. **576**, 217 (2005).
9. 1 V. Mikhelashvili, G.Eisenstein, and F. Edelmann, J. Appl. Phys. **90**, 5447 (2001).
10. V.A. Rozhkov, A.Y. Trusova, and I.G. Berezhnoy, Thin Solid Films **325**, 151 (1998).
11. P. Delugas and V. Fiorentini, Microelectronics Reliability **45**, 831 (2005).
12. G. Seguini, E. Bonera, S. Spiga, G. Scarel, and M. Fanciulli, Appl. Phys. Lett., **85**, 5316 (2004).
13. W. Cai, S. E. Stone, J. P. Pelz, L. F. Edge and D. G. Schlom, Appl. Phys. Lett. **91**, 042901 (2007).
14. J. Kwo, M. Hong, A.R. Kortan, K.L. Queeny, Y.J. Chabal, R.L. Opila, D.A. Müller, S.N.G. Chu, J. Appl. Phys. **89**, 3920 (2001).
15. A. Fissel, H.J Osten, and E. Bugiel, J. Vac. Sci. Technol. B **21**, 1765 (2003).
16. H. J. Osten, E. Bugiel, M. Czernohorsky, Z. Elassar, O. Kirfel, and A. Fissel, Topics in Appl. Phys, **106**, 101 (2007).
17. A. Laha, H.J. Osten, and A. Fissel, Appl. Phys. Lett. **89**, 143514 (2006).
18. A. Fissel, Z. Elassar, E. Bugiel, M. Czernohorsky, O. Kirfel and H. J. Osten, J. Appl. Phys. **99**, 074105 (2006).
19. D. Schmeisser, J. Dabrowski, H.-J. Muessig, Mater. Sci. Engin. B **109**, 30 (2004).
20. M. Czernohorsky, A. Fissel, E. Bugiel, O.Kirfel, and H.J. Osten, Appl. Phys. Lett. **88**, 152905 (2006).
21. A. Laha, H.J. Osten, and A. Fissel, Appl. Phys. Lett. **90**, 113508 (2007).
22. M.C. Lemme, H.D.B. Gottlob, T.J. Echtermeyer, H. Kurz, R. Endres, U. Schwalke, M. Czernohorsky, and H.J. Osten, J. Vac. Sci. & Technol. B**27**, 258 (2009).
23. T. Echtermeyer, H.D.B. Gottlob, T. Wahlbrink, T. Mollenhauer, M. Schmidt, J.K. Efavi, M.C. Lemme, and H. Kurz, Solid-State Electronics **51**, 617 (2007).
24. R. Endres, Y. Stefanov, and U. Schwalke, Microelectron. Reliab. **47**, 528 (2007).
25. Q.-Q. Sun, A. Laha, S.-J. Ding, D. W. Zhang, H. J. Osten, and A. Fissel: Appl. Phys. Lett. **92**, 152908 (2008).
26. M. Czernohorsky, D. Tetzlaff, E. Bugiel, R. Dargis, H.J. Osten, H. D. B. Gottlob, M. Schmidt, M. C. Lemme, and H. Kurz, Semicond. Sci. & Technol. **23**, 035010 (2008).
27. J. Kwo, M. Hong, A. R. Kortan, K. T. Queeney, Y. J. Chabal, J. P. Mannaerts, T. Boone, J. J. Krajewski, A. M. Sergent, and J. M. Rosamilia, Appl. Phys. Lett. **77**, 130 (2000).
28. H. Kroemer, in *Heteroepitaxy on Si,* MRS Symposia Proceedings No. 67, edited by J. C. C. Fan and J. M. Poate (Materials Research Society, Pittsburgh, PA, 1986), and references therein.
29. A. Laha, E. Bugiel, J.X. Wang, Q.Q. Sun, A. Fissel, and H.J. Osten, Appl. Phys. Lett. **93**, 182907 (2008).
30. H.J. Osten, J.P. Liu, E. Bugiel, H.J. Müssig, and P. Zaumseil, Mat Sci. & Engineer. B **87** (2001) 297.
31. A. Laha, E. Bugiel, R. Dargis, D. Schwendt, M. Badylevich, V.V. Afanas'ev, A. Stesmans, A. Fissel, and H.J. Osten, Microelectronic Journal **40** (2009) 633.

Alternate Channel Materials

Mater. Res. Soc. Symp. Proc. Vol. 1155 © 2009 Materials Research Society   1155-C13-02

Chemical beam deposition of high-k gate dielectrics on III-V semiconductors: TiO$_2$ on In$_{0.53}$Ga$_{0.47}$As

Roman Engel-Herbert[1], Yoontae Hwang[1], James M. Lebeau[1], Yan Zheng[2], and Susanne Stemmer[1]
[1]Materials Department, University of California, Santa Barbara, CA 93106-5050, U.S.A.
[2]Department of Electrical and Computer Engineering, University of California, Santa Barbara, CA 93106-5050, U.S.A.

ABSTRACT

We report on the growth of high-permittivity (k) TiO$_2$ thin films on In$_{0.53}$Ga$_{0.47}$As channels by chemical beam deposition with titanium isopropoxide as the source. The films grew in a reaction-limited regime with smooth surfaces. High-resolution transmission electron microscopy showed an atomically abrupt interface with the In$_{0.53}$Ga$_{0.47}$As channel that indicated that this interface is thermally stable. Measurements of the leakage currents using metal-oxide-semiconductor capacitors with Pt top electrodes revealed asymmetric characteristics with respect to the bias polarity, suggesting an unfavorable band alignment for CMOS applications. X-ray photoelectron spectroscopy was used to determine the TiO$_2$/In$_{0.53}$Ga$_{0.47}$As band offsets. A valence band offset of 2.5 ± 0.1 eV was measured.

INTRODUCTION

The scaling of conventional, Si-based complementary metal oxide semiconductor (CMOS) field-effect-transistors (FETs) is approaching its fundamental limits. High-mobility semiconductor channel materials, such as III-V compound semiconductors or Ge, have the potential to enable further scaling of FETs, thus allowing for higher speed and lower operating voltages. The realization of III-V based MOSFETs requires the development of suitable gate dielectrics. These dielectrics should be thermally stable in contact with the semiconductor and should allow for a low density of interface states and low leakage currents, which require sufficient band offsets with the semiconductor. For scaling, a high dielectric constant is also necessary. One of the main challenges in III-V CMOS is that oxygen exposure of the semiconductor surface must be prevented because native III-V oxides can cause Fermi level pinning even at sub-monolayer coverage [1].

Several binary oxides, such as Al$_2$O$_3$ [2,3], ZrO$_2$ [4] and HfO$_2$ [5,6], have been investigated for use as high-k dielectrics on InGaAs channels, mainly by atomic layer deposition (ALD). This deposition technique combines several advantages, such as low temperature, large area uniformity and precise thickness control. Ideally, however, a growth environment is desired where the dielectric is deposited on a well-defined, clean III-V surface without any oxidation of the semiconductor during deposition. This would allow for distinguishing between the intrinsic properties of different high-k/III-V combinations and those arising from interface degradation during gate stack deposition, such as unintentional oxidation.

One way to achieve this is to use chemical beam deposition in ultrahigh vacuum on III-V surfaces that have never been exposed to air, which can be obtained by As-capping after growth and de-capping in vacuum immediately before high-k deposition. Of particular interest are metal oxide precursors for which the metal ion already comes bonded to oxygen [7]. In this case, no additional oxidant is needed during growth, thereby reducing the possibility of oxidation of the semiconductor surface. In addition, the ultrahigh vacuum environment enables characterization of surfaces and growth with in-situ monitoring techniques, such as reflection high-energy electron diffraction (RHEED).

Here, we report on the growth, electrical characterization and band offsets of TiO_2 on $In_{0.53}Ga_{0.47}As$ channels. TiO_2 is attractive, because it has one of the largest dielectric constants of all binary oxides [8]. The TiO_2 was deposited in a molecular beam epitaxy (MBE) system using low-pressure chemical beam deposition. High-purity titanium isopropoxide $Ti(Oi-Pr)_4$ (99.999% Sigma Aldrich, USA) was used as metal organic precursor. Its high volatility allows for thermal evaporation at relatively low temperatures and high growth rates [9].

EXPERIMENT

Substrates were As-capped 200 nm thick n-doped (Si: 1×10^{17} cm^{-3}) $In_{0.53}Ga_{0.47}As$ layers grown lattice matched on (001) oriented n+InP wafers (S: ~10^{19} cm^{-3}) and were supplied by a commercial vendor (IntelliEpi). Before TiO_2 growth, the As-oxide on top of the As-cap was desorbed for at least 20 min at a substrate temperature of 200 °C (thermocouple reading) [10]. The temperature was subsequently ramped to 250 °C to thermally remove the As-cap in the absence of As flux, monitored by RHEED. The substrate temperature was subsequently increased to the growth temperature of 300 °C to promote the thermolysis of $Ti(O^i-Pr)_4$. The precursor was supplied to the growth chamber (base pressure ~ 8×10^{-9} torr) through a heated gas inlet system, equipped with a pressure controlled leak valve. No carrier gas or additional oxidant was used. The precursor flux corresponded to a beam equivalent pressure (BEP) of 5.0×10^{-6} torr, measured at the sample position. The growth rate was 0.5 nm/min.

Metal-oxide-semiconductor capacitor (MOSCAP) structures were fabricated for electrical characterization. The back ohmic contacts were electron-beam deposited Ni/AuGe/Ni/Au layer stacks which were annealed at 350 °C for 1 min in oxygen. Pt and Al (150 nm thick) were used as top contacts and were evaporated through a shadow mask and annealed at 500 °C for 1 min. Current-voltage characteristics were measured with a semiconductor parameter analyzer (HP 4155 B, Agilent Technologies). Atomic force microscopy was used to investigate the surface morphology. High-resolution transmission electron microscopy (HRTEM) and high-angle annular dark-field imaging in scanning transmission electron microscopy (HAADF-STEM) studies were carried out using a 300 kV field-emission microscope (FEI Titan 80-300 TEM/STEM). X-ray photoelectron spectroscopy (XPS) was performed on 5 nm and 20 nm thick TiO_2 films, respectively. An InGaAs reference sample was prepared in-situ in the XPS system (Kratos Axis Ultra) with the same oxide desorption and As de-capping procedure as described above. Survey and high-resolution spectra were recorded at a pass energy of 80 and 20 eV and a step size of 0.5 and 0.05 eV, respectively. Monochromated Al K_α 1486.6 eV x-ray radiation and a source power of 270 W was used. Charge correction was performed for the sample with the 5 nm thick TiO_2 film by shifting the spectra such that the C 1s peak was at the known position of adventitious carbon with the binding energy of 285.0 eV. The core level peaks Ti $2p_{3/2}$ and In $3d_{5/2}$ were used as reference peak positions to align the XPS spectra of the InGaAs reference sample and 20 nm thick TiO_2 film with respect to the 5 nm thick TiO_2/InGaAs sample.

RESULTS AND DISCUSSION

Interface structure

Figure 1 shows atomic-resolution HAADF-STEM and HRTEM cross-section images of the $TiO_2/In_{0.53}Ga_{0.47}As$ interface for a TiO_2 film grown at 300 °C. A direct interface is observed between crystalline TiO_2 and the $In_{0.53}Ga_{0.47}As$ channel. In particular, the chemical sensitivity of HAADF-STEM images confirms the absence of any amorphous interface layers, such as the native InGaAs oxides. This showed that the deposition method is suitable to avoid any unintentional oxidation of the semiconductor. Furthermore, in sharp contrast to Si [11], TiO_2 appears to be thermally stable on $In_{0.53}Ga_{0.47}As$.

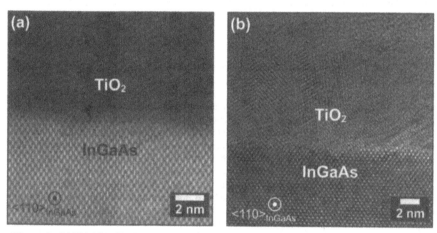

Figure 1. (a) HAADF-STEM and (b) HRTEM cross-section images of the $TiO_2/In_{0.53}Ga_{0.47}As$ interface recorded along $<110>_{InGaAs}$.

Surface morphology

Figure 2 shows the surface morphology of a 10 nm TiO_2 film as measured by AFM in tapping mode. The films were very smooth having a root mean square (rms) roughness of only 0.65 nm, indicating good wetting behavior.

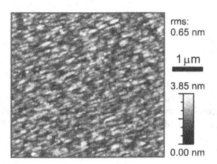

rms:
0.65 nm

1 μm

3.85 nm

0.00 nm

Figure 2. AFM image of the surface of a 10 nm thick TiO_2 film on $In_{0.53}Ga_{0.47}As$.

Leakage currents

Figure 3 shows the leakage current densities for MOSCAPs with 10 nm thick TiO_2 films and two different top electrodes, Al and Pt. The leakage current for the MOSCAP with the Pt top electrode is relatively high (~ 0.1 A/cm^2 at gate bias of -1 V) and is slightly asymmetric. The asymmetry of the curve with Pt top electrode suggests different conduction and valence band offsets with $In_{0.53}Ga_{0.47}As$. The higher leakage current in accumulation indicates a smaller conduction band offset. The high leakage currents may also be due to oxygen deficiency, giving rise to n-type conductivity. TiO_2 films grown from this precursor without additional oxidant are oxygen deficient [9]. For the MOSCAP with the Al top contact the leakage current density is strongly reduced by more than three orders of magnitude compared to Pt and the curve is symmetric in forward and reverse bias. The reduction of the leakage with the Al electrode is attributed to the formation of an aluminum oxide layer at the electrode interface, which is known to form when Al is in contact with high-k dielectrics [12]. This also explains the symmetric shape, as the leakage properties are completely governed by the blocking aluminum oxide layer.

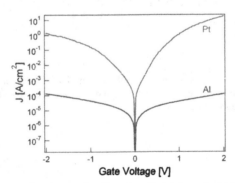

Figure 3. Leakage current density as a function of applied voltage for MOSCAPs with Al and Pt top contacts, respectively.

Band offset measurements by XPS

XPS measurements were performed to quantitatively determine the band offsets between TiO_2 and $In_{0.53}Ga_{0.47}As$ using the method by Kraut et al. [13]. Figure 4 shows a schematic band diagram, with alignments as determined below. The valence band offset ΔE_{VB} is given by:

$$\Delta E_{VB} = \left(E_{In\,3d\,5/2} - E_{VB}\right)_{InGaAs} - \left(E_{Ti\,2p\,3/2} - E_{VB}\right)_{TiO_2} + \left(\Delta E_{CL}\right)_{TiO_2/InGaAs}, \qquad (1)$$

where $E_{Ti\,2p\,3/2}$ and $E_{In\,3d\,5/2}$ are the core level binding energies and E_{VB} are the valence band maxima of $In_{0.53}Ga_{0.47}As$ and TiO_2. The valence band maxima were obtained by linear fits of the valence band edges and extrapolating to the zero line, as shown in Fig. 5(b). The In 3d core level were fit using the parameters given in ref. [14] to determine the exact binding energy. From high resolution spectra of the $In_{0.53}Ga_{0.47}As$ reference sample and the 20 nm thick TiO_2 the energy difference between core level and valence band maximum was determined to 443.8 ± 0.1 eV and 456.0 ± 0.1 eV, respectively.

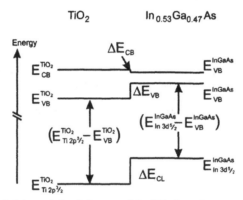

Figure 4. Schematic band diagram of the $TiO_2/In_{0.53}Ga_{0.47}As$ interface.

The core level offset $\Delta E_{CL} = E_{Ti\,2p\,3/2} - E_{In\,3d\,5/2}$ was obtained from the analysis of the $TiO_2/InGaAs$ heterojunction sample (5 nm thick TiO_2 film on InGaAs), shown in Fig. 5(a). With a core level offset of 14.7 ± 0.1 eV the valence band offset was determined to 2.5 eV. Aligning the valence band spectra using the core level peaks Ti $2p_{3/2}$ and In $3d_{5/2}$ with respect to the 5 nm thick $TiO_2/InGaAs$ sample allows for a direct comparison of the valence band edges, shown in Fig 5(b). The valence band spectra are well aligned and the individual valence band features of TiO_2 and InGaAs can be distinguished in spectra of the 5 nm $TiO_2/In_{0.53}Ga_{0.47}As$ sample (black curve). The valence band offset between InGaAs and TiO_2 is indicated. In conjunction with the values for band gap of $In_{0.53}Ga_{0.47}As$ (0.75 eV [15]) and that of TiO_2 (~ 3.2 eV [16]), these values show that there is almost no conduction band offset at this interface within the accuracy of XPS.

115

The high-resolution spectrum in Fig. 5(a) also reveals that Ti is present in a lower oxidation state than 4+ in the film. The slightly increased intensities on the lower binding energy sides of Ti $2p_{3/2}$ and Ti $2p_{1/2}$ peaks are attributed to Ti with a lower oxidation state. This indicates that oxygen vacancies might be present in the TiO_2 film, which are compensated by a lower Ti valence state.

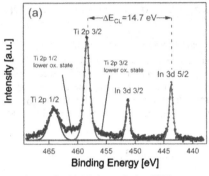 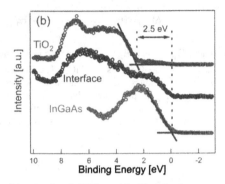

Figure 5. (a) High resolution XPS spectrum of the core levels Ti 2p and In 3d of the 5 nm $TiO_2/In_{0.53}Ga_{0.47}As$ sample. The tails at the lower binding energy side of Ti $2p_{3/2}$ and Ti $2p_{1/2}$ indicate that Ti is present with an oxidation state of less than 4+. (b) Valence band spectra of a 20 nm thick TiO_2 film (blue) and InGaAs (red) aligned with respect to the 5 nm thick TiO_2 film (black). Linear extrapolation of the valence band edges is shown which allow for a determination of the magnitude of the valence band offset ($\sim$ 2.5 eV).

CONCLUSIONS

The results demonstrate the excellent potential of chemical beam deposition for the integration of high-k gate dielectrics on III-V semiconductors with abrupt interfaces free of the native oxides of the III-V semiconductor. Using this method, smooth TiO_2 films with atomically abrupt interfaces were fabricated on $In_{0.47}Ga_{0.53}As$ channels. Measurements of leakage currents and band offsets indicate insufficient conduction band offsets at the $TiO_2/InGaAs$ interface.

ACKNOWLEDGMENTS

The authors acknowledge funding from the Semiconductor Research Corporation under the Nonclassical CMOS Research Center (Task 1437.003) and Intel Custom Funding (Task 1635.002). One author (R. E.-H.) also acknowledges support from the Alexander-von-Humboldt Foundation through a Feodor-Lynen fellowship. This work made use of the UCSB Nanofabrication Facility, a part of the NSF-funded NNIN network.

REFERENCES

[1]   W. E. Spicer, P. W. Chye, P. R. Skeath, C. Y. Su, and I. Lindau, J. Vac. Sci. Techn. **16**, 1422 (1979).

[2]   J. P. d. Souza, E. Kiewra, Y. Sun, A. Callegari, D. K. Sadana, G. Shahidi, D. J. Webb, J. Fompeyrine, R. Germann, C. Rossel, and C. Marchiori, Appl. Phys. Lett. **92**, 153508 (2008).

[3]   H. C. Chiu, L. T. Tung, Y. H. Chang, Y. J. Lee, C. C. Chang, J. Kwo, and M. Hong, Appl. Phys. Lett. **93**, 202903 (2008).

[4]   S. Koveshnikov, N. Goel, P. Majhi, H. Wen, M. B. Santos, S. Oktyabrsky, V. Tokranov, R. Kambhampati, R. Moore, F. Zhu, J. Lee, and W. Tsai, Appl. Phys. Lett. **92**, 222904 (2008).

[5]   Y. C. Chang, M. L. Huang, K. Y. Lee, Y. J. Lee, T. D. Lin, M. Hong, J. Kwo, T. S. Lay, C. C. Liao, and K. Y. Cheng, Appl. Phys. Lett. **92**, 072901 (2008).

[6]   H.-S. Kim, I. Ok, F. Zhu, M. Zhang, S. Park, J. Yum, H. Zhao, P. Majhi, D. I. Garcia-Gutierrez, N. Goel, W. Tsai, C. K. Gaspe, M. B. Santos, and J. C. Lee, Appl. Phys. Let. **93**, 132902 (2008).

[7]   R. C. Smith, T. Z. Ma, N. Hoilien, L. Y. Tsung, M. J. Bevan, L. Colombo, J. Roberts, S. A. Campbell, and W. L. Gladfelter, Adv. Mater. Opt. Electron. **10**, 105 (2000).

[8]   R. D. Shannon, J. Appl. Phys. **73**, 348 (1993).

[9]   B. Jalan, R. Engel-Herbert, J. Cagnon, and S. Stemmer, J. Vac. Sci.Technol. A **27**, 230 (2009).

[10]   U. Resch, N. Esser, Y. S. Raptis, W. Richter, J. Wasserfall, A. Förster, and D. I. Westwood, Surf. Sci. **269-270**, 797 (1992).

[11]   K. J. Hubbard and D. G. Schlom, J. Mater. Res. **11**, 2757 (1996).

[12]   H. Kim, P. C. McIntyre, C. O. Chui, K. C. Saraswat, and S. Stemmer, J. Appl. Phys. **96**, 3467 (2004).

[13]   E. A. Kraut, R. W. Grant, J. R. Waldrop, and S. P. Kowalczyk, Phys. Rev. B **28**, 1965 (1983).

[14]   M. Procop, J. Electron Spectrosc. & Rel. Phenom. **59**, R1 (1992).

[15]   K. H. Goetz, D. Bimberg, H. Jurgensen, J. Selders, A. V. Solomonov, G. F. Glinskii, and M. Razeghi, J. Appl. Phys. **54**, 4543 (1983).

[16]   L. Kavan, M. Gratzel, S. E. Gilbert, C. Klemenz, and H. J. Scheel, J. Am. Chem. Soc. **118**, 6716 (1996).

Mater. Res. Soc. Symp. Proc. Vol. 1155 © 2009 Materials Research Society   1155-C10-03

Atomic Layer Deposition of Metal Oxide Films on GaAs (100) Surfaces

Theodosia Gougousi,[1] John W. Lacis,[1] Justin C. Hackley,[1] and J. Derek Demaree[2]
[1] Department of Physics, University of Maryland Baltimore County (UMBC), Baltimore, MD 21250 U.S.A.
[2] Weapons & Materials Research Directorate, Army Research Laboratory, Aberdeen Proving Ground, MD 21005-5069 U.S.A.

ABSTRACT

Atomic Layer Deposition is used to deposit HfO_2 and TiO_2 films on GaAs (100) native oxides and etched surfaces. For the deposition of HfO_2 films two different but similar ALD chemistries are used: i) tetrakis dimethyl amido hafnium (TDMAHf) and H_2O at 275°C and ii) tetrakis ethylmethyl amido hafnium (TEMAHf) and H_2O at 250°C. TiO_2 films are deposited from tetrakis dimethyl amido titanium (TDMATi) and H_2O at 200°C. Rutherford Back Scattering shows linear film growth for all processes. The film/substrate interface is examined using x-ray photoelectron spectroscopy and confirms the presence of an "interfacial cleaning" mechanism.

INTRODUCTION

The deposition of high dielectric constant (high-k) films on Si surfaces has been studied extensively as a means to extend the lifetime of Si-based microelectronics. One of the major issues in the integration of high-k materials with Si is the inadvertent formation of SiO_2 interfacial layers even for oxides that are predicted to be thermodynamically stable on Si surfaces.[1-2] High mobility substrates such as GaAs and InGaAs have better electrical properties than Si but their use in the semiconductor industry has been hindered by the poor electrical quality of their native oxides.[3] However, high-k dielectrics can be deposited on any semiconductor surface and there are several recent reports of good quality devices using high-ks on GaAs surfaces.[4] Most interestingly, there are several reports of an interface etching reaction taking placing during the ALD of HfO_2 and Al_2O_3 on native oxide covered GaAs surfaces that results in a very thin interfacial layer between the high-k film and the GaAs substrate.[5-12]The common thread of all these observations is the use of metal organic precursors as the metal source. In this article we examine the evolution of the high-k/GaAs interface for three ALD processes that use metal organic precursors of the amide family.

EXPERIMENT

Film depositions were performed in a hot-wall reactor described elsewhere.[13] HfO_2 films were deposited using two different metal organic amide precursors. i) tetrakis dimethyl amido hafnium (TDMAHf) and H_2O at 275°C and ii) tetrakis ethylmethyl amido hafnium (TEMAHf) and H_2O at 250°C. For the deposition of TiO_2 films tetrakis dimethyl amido titanium (TDMATi) and H_2O were used at 200°C. Three different starting surfaces are examined. Surfaces termed "native oxide" were prepared by cleaning pieces of GaAs (100) wafers in acetone, methanol, rinsing in deionized (DI) water and blown dry in a N_2 stream. This preparation has been shown to preserve the surface native oxides.[11] Surfaces with very little oxide coverage can be prepared by: 5 min soak in JT Baker 100 (JTB) solution, 5 min rinse in DI water, and then by either etching for 30 s in Buffered Oxide Etch (BOE) and quick DI and N_2 blow dry ("HF" surfaces) or etching for 3 min in 30 % aqueous NH_4OH solution ("NH4OH" surfaces). Si control samples were prepared by soaking pieces of native oxide covered Si (100) wafers in JTB for 5 min, followed by 5 min DI rinse and N_2 blow dry.

Ex-situ x-ray photoelectron spectroscopy (XPS) was used to examine the composition of the interface and performed with a Kratos AXIS 165 (Al x-ray source, 1486.6 eV), equipped with a hemispherical analyzer (165 mm radius). The high resolution spectra were baseline corrected using Shirley backgrounds and deconvolved using Gaussian-Lorentzian functions. The substrate As $3d_{5/2}$ and $_{3/2}$ doublet was deconvolved by assuming functions of equal Full width at Half Maximum (FWHM), a spin-orbit separation of 0.7 eV and intensity ratio of 3:2. Sample charging effects were corrected by placing the substrate As $3d_{5/2}$ peak (As-Ga) at a binding energy (BE) of 41.1 eV and shifting the rest of the regions accordingly.[14] For the contribution of the various arsenic oxides (As_2O_3, As_2O_5, AsO_x) in the 3d region a single function was found sufficient due to the well resolved substrate and oxide peaks. Due to the complexity of the Ga $2p_{3/2}$ region the spectra were fitted using 2 peaks, one corresponding to the substrate peak (Ga-As), and another representing the total contribution of the gallium oxides.

Rutherford backscattering spectrometry (RBS) measurements were made using a 1.2 MeV He^+ beam obtained from a National Electrostatics 5SDH-2 positive ion accelerator. The backscattering angle was 170 degrees, and the spectra were collected using a surface barrier detector subtending approximately 5 millisteradians. The raw RBS data was fitted using the simulation program RUMP.[15]

RESULTS

TEMAHf + H2O

Figure 1 shows high resolution XP spectra for the As 3d and Ga $2p_{3/2}$ regions for the starting surface and after 25 ALD cycles that result in the deposition of ~30Å of HfO_2 film. The starting surface is cleaned GaAs (100) covered with ~26Å of native oxide. The presence of the HfO_2 layer in sample (b) reduces the overall intensity of both the substrate and the oxides peaks. However the ratio of the integrated oxide intensity to the integrated substrate As 3d or Ga $2p_{3/2}$ intensity is independent of the overlayer thickness as the photoelectrons from both regions are attenuated by the same exponential factor when crossing the HfO_2 layer. For the As 3d

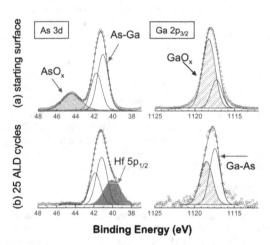

Binding Energy (eV)

Figure 1. As 3d and Ga $2p_{3/2}$ high resolution XP spectra for (a) the starting native oxide GaAs (100) surface, and (B) after the deposition of ~30Å of HfO_2 from TEMAHf and H_2O.

120

region the ratio of the oxide to the substrate peak decreases from 0.31±0.04 for the starting surface to 0.03±0.01 after the deposition of ~30Å of HfO$_2$.This reduction is clearly evident in comparing spectra (a) and (b). The very small separation in the binding energy of the Ga substrate peak and that of the oxides makes direct observation of the Ga oxide intensity reduction harder. However, upon closer inspection clear differences are seen in the shape of the peak between samples (a) and (b). Peak (a) exhibits a low BE tail while peak (b) exhibits a high BE tail indicating a change in the mixture of the constituent peaks. Peak deconvolution reveals that the ratio of the integrated Ga oxide intensity to that of the substrate changes from 2.9±0.4 for the starting surface to 0.9±0.1 after 25 ALD cycles. The validity of the parameters used in the peak deconvolution especially for the Ga 2p$_{3/2}$ is demonstrated in Figure 2 that shows the same spectral regions for 20 cycle films (~25Å of HfO2) deposited on etched GaAs surfaces. Both treatments ("HF" and "NH$_4$OH") have been shown to remove most of the surface native oxides.[11] After the deposition of a small amount of interfacial oxide is detected but the FWHM of the Ga 2p$_{3/2}$ peak is clearly smaller providing the essential parameters to deconvolve the rest of the Ga 2p$_{3/2}$ peaks.

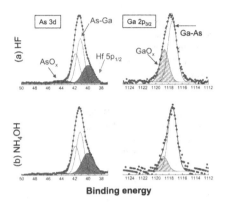

Figure 2. As 3d and Ga 2p$_{3/2}$ spectra for samples of ~25Å of HfO$_2$ deposited on GaAs surfaces treated in HF and NH$_4$OH solutions..

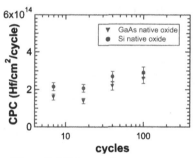

Figure 3. Hf atom surface coverage per cycle for cleaned native oxide GaAs (100) surfaces and Si native oxide control samples.

Figure 3 shows Hf atom coverage per cycle measured by RBS for films deposited simultaneously on GaAs cleaned native oxide and Si control samples. Similar to previous observations the Hf atom coverage per cycle (CPC) is initially low but appears to reach a steady state value of ~2.7x10^{14} cm^{-2} after ~40 process cycles.

TDMAHf + H$_2$O

Figure 4 shows high resolution XP spectra for the As 3d and Ga $2p_{3/2}$ regions for 15, 20 and 100 cycle films deposited on native oxide starting surfaces similar to those shown on Figure 1 (a). The nominal HfO_2 thickness of these samples is 15, 20 and 100 Å.[16] For the 100 cycle film the HfO_2 layer was ion etched in the XPS chamber until the substrate peaks were clearly visible. As evidenced by the presence of the Hf $5p_{1/2}$ peak some of the HfO_2 layer was left to preserve the integrity of the interface. The sequence of the data shows a gradual removal of the interfacial Ga and As oxides. As oxides are removed easier; there is practically zero intensity for the 20 cycle film. Ga oxides persist longer and even after 100 ALD cycles a monolayer or two is still present in the interface.

Figure 5 shows Hf atom coverage per cycle (CPC) measured by RBS for films deposited on GaAs native oxide, "HF" and "NH₄OH" surfaces. The coverage reached reaches a steady state value of ~2.9×10^{14} cm^{-2} for all three starting surfaces after ~20 process cycles.

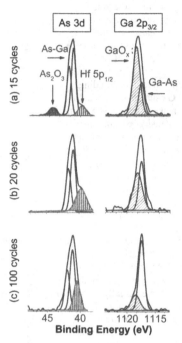

Figure 4. As 3d and Ga $2p_{3/2}$ high resolution XP spectra for 15, 20, and 100 cycle HfO2 films deposited on native oxide GaAs (100) surfaces from TDMAHf and H_2O.

TDMATi + H₂O

Figure 6 shows high resolution XP spectra for the As 3d and Ga $2p_{3/2}$ regions for the starting surface and after 60 ALD cycles that result in the deposition of ~35Å of TiO_2 film. For the As 3d region the ratio of the oxide to the substrate peak decreases from 0.53±0.08 for the starting surface to 0.19±0.03 for the ~35Å TiO_2 sample. This reduction is clearly evident in comparing spectra (a) and (b). For

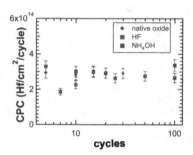

Figure 5. Hf atom surface coverage per cycle on various GaAs surfaces measured by RBS.

122

the Ga 2p 3/2 peak, deconvolution reveals that the ratio of the integrated Ga oxide intensity to that of the substrate changes from 3.4±0.5 for the starting surface to 1.1±0.2 after 60 ALD cycles.

Figure 7 shows Ti atom coverage per cycle (CPC) measured by RBS for films deposited on Si native oxide surfaces. The coverage is has a steady state value of ~1.4 x10^{14} cm^{-2}. The Ti atom surface coverage on GaAs surfaces can not be measured by RBS. However, based on the results for the ALD of HfO$_2$ on GaAs we do not expect significant difference in surface coverage between the Si native oxide and GaAs native oxide.

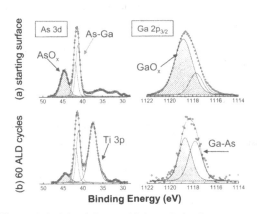

Figure 6. As 3d and Ga 2p$_{3/2}$ high resolution XP spectra for (a) the starting native oxide GaAs surface, and (b) after the deposition of ~40Å of TiO$_2$ from TDMATi and H$_2$O.

DISCUSSION

When native oxide GaAs substrates are used as a starting surface, the ALD of HfO$_2$ and TiO$_2$ results in the gradual thinning of the surface As and Ga oxides. Arsenic oxides are easier to remove while a monolayer or two of the gallium oxides remains at the interface even after 100 Å of HfO$_2$ has been deposited. The common feature behind these observations is the use of amide metal organic precursors. Two of the precursors (TDMAHf and TDMATi) have exactly the same structure while the third (TEMAHf) is a common variant in which one of the methyl ligands has been replaced by an ethyl ligand. It appears that one of the ALD reaction byproducts, an amine HNR$_1$R$_2$ where R$_1$ and R$_2$ are either ethyl of methyl groups may be responsible for the etching reaction. Evidence in support of this are:

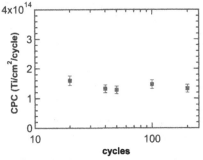

Figure 7. Ti atom coverage per cycle on Si native oxide surfaces measured by RBS.

a. A gradual reduction in the oxide intensity (relative to the substrate intensity) as a function of film thickness is observed for coalesced TiO$_2$ and HfO$_2$ films. High-k films coalescence is required to prevent postdeposition oxidation of the interface.

b. For the three ALD processes examined in this work, the process that results in the lowest film growth rate requires more ALD cycles to complete the oxide removal. The TiO$_2$ process that achieves the lowest metal surface coverage requires in excess of 60 ALD cycles to reach ~60% decrease in the As oxide signal intensity. By contrast the TDMAHf process that has a substantially higher growth rate requires only about 20 ALD cycles to reach almost complete removal of the As oxide.

Further supporting evidence can be found in the reports of surface oxide removal caused by the thermal decomposition of metal organic molecules such as tris dimethyl amino arsenic on GaAs surfaces.[17-18]

ACKNOWLEDGMENTS

Acknowledgment is made to the Donors of the American Chemical Society Petroleum Research Fund and the UMBC ADVANCE Institutional Transformation Program (NSF-0244880) for partial support of this research. We would like to thank Karen G. Gaskell for collecting some of the XPS data.

REFERENCES

1. G.D. Wilk, R.M. Wallace, J.M. Anthony, J. Appl. Phys. **89**, 5243 (2001).
2 R.M.C. de Almeida and I.J.R. Baumvol, Surface Science Reports 49, 1, (2003).
3 J. Robertson and B. Falabretti, J. Appl. Phys. **100**, 014111 (2006).
4 H.C. Lin, T. Yang, H. Sharifi, S.K. Kim, Y. Xuan, T. Shen, S. Mohammadi, and P.D. Ye Appl. Phys. Lett. 91 (21) 212101 (2007).
5 P. D. Ye, G. D. Wilk, B. Yang, J. Kwo, S. N. G. Chu, S. Nakahara, H.-J. L. Gossmann, J. P. Mannaerts, M. Hong, K. K. Ng, and J. Bude Appl. Phys. Lett. **83**, 180 (2003).
6 M. M. Frank, G. D. Wilk, D. Starodub, T. Gustafsson, E. Garfunkel, .Y. J. Chabal J. Grazul and D. A. Muller Appl. Phys. Lett. **86**, 152904 (2005).
7 M. L. Huang, Y. C. Chang, C. H. Chang, Y. J. Lee, P. Chang, J. Kwo,T. B. Wu and M. Hong, Appl. Phys. Lett. **87**, 252104 (2005).
8 C.-H. Chang, Y.-K. Chiou, Y.-C. Chang, K.-Y. Lee, T.-D. Lin, T.-B. Wu, M. Hong, and J. Kwo, Appl. Phys. Lett. **89**, 242911 (2006).
9 C. L. Hinkle, A. M. Sonnet, E. M. Vogel, S. McDonnell, G. J. Hughes, M. Milojevic, B. Lee, F. S. Aguirre-Tostado, K. J. Choi, H. C. Kim, J. Kim, and R. M. Wallace, Appl. Phys. Lett. **92**, 071901 (2008).
10 Y. C. Chang, M. L. Huang, K. Y. Lee, Y. J. Lee, T. D. Lin, M. Hong, J. Kwo, T. S. Lay, C. C. Liao, and K. Y. Cheng, Appl. Phys. Lett. **92**, 072901 (2008).
11 J.C. Hackley, J.D. Demaree, and T. Gougousi, Appl. Phys. Lett. **92**(16), 162902 (2008).
12 D. Shahrjerdi, D. I. Garcia-Gutierrez, E. Tutuc, and S. K. Banerjee, Appl. Phys. Lett. **92** (22), 223501, (2008).
13 J.C. Hackley, T. Gougousi, J.D. Demaree, J. Appl. Phys. **102**, 034101 (2007).
14. C.C. Surdu-Bob, S.O. Saied, J.L. Sullivan, Appl. Surf. Sci. **183**, 126 (2001).
15. L. R. Doolittle, Nucl. Instrum. Meth. **B15**, 227 (1986).
16 J.C. Hackley, J.D. Demaree, and T. Gougousi, J. Vac. Sci. Technol. A **25**(5), 1235 (2008).
17 D. Marx, H. Asahi X.F. Liu, M. Higashiwaki, A.B. Villaflor, K. Miki,K. Yamamoto, S. Gonda, S. Shimomura, S. Hiyamizu J. Crystal Growth **150**, 155 (1995).
18 H. Asahi X.F. Liu, K. Inoue, D. Marx, K. Asami, K. Miki, S. Gonda, J. Crystal Growth **145**, 688 (1994).

Mater. Res. Soc. Symp. Proc. Vol. 1155 © 2009 Materials Research Society   1155-C02-03

Electron Scattering in Buried InGaAs MOSFET Channel With HfO$_2$ Gate Oxide

S. Oktyabrsky,[1] P. Nagaiah,[1] V. Tokranov,[1] S. Koveshnikov,[1,2] M. Yakimov,[1] R. Kambhampati,[1] R. Moore,[1] and W. Tsai[2]

[1]College of Nanoscale Science and Engineering, University at Albany-SUNY, Albany, NY 12203, U.S.A.
[2]Intel Corporation, Santa Clara, 95052, U.S.A.

ABSTRACT

Group III-V semiconductor materials are being studied as potential replacements for conventional CMOS technology due to their better electron transport properties. However, the excess scattering of carriers in MOSFET channel due to high-k gate oxide interface significantly depreciates the benefits of III-V high-mobility channel materials. We present results on Hall electron mobility of buried QW structures influenced by remote scattering due to InGaAs/HfO$_2$ interface. Mobility in In$_{0.77}$Ga$_{0.23}$As QWs degraded from 12000 to 1200 cm^2/V-s and the mobility vs. temperature slope changed from T$^{-1.2}$ to almost T$^{+1.0}$ in 77-300 K range when the barrier thickness is reduced from 50 to 0 nm. This mobility change is attributed to remote Coulomb scattering due to charges and dipoles at semiconductor/oxide interface. Elimination of the InGaAs/HfO$_2$ interface via introduction of SiO$_x$ interface layer formed by oxidation of thin a-Si passivation layer was found to improve the channel mobility. The mobility vs. sheet carrier density shows the maximum close to 2x10^{12} cm^{-2}.

INTRODUCTION

As CMOS scaling continues, new channel materials including III-V compound semiconductors would be needed to improve channel transport, intrinsic gate delay and to reduce operating supply voltage of MOSFETs [1]. In order to keep high mobility in the channel, it has to be separated from the primary scattering sources, and thus buried quantum well (QW) channel is a promising option, although the top barrier adds up to the equivalent oxide thickness. It is, therefore, important to optimize the design of the gate dielectric consisting of a wide-bandgap semiconductor top barrier and high-k gate oxide layers. This paper is the first attempt to understand how significantly the top barrier layer reduces the scattering associated with the high-k oxide.

The mobility is typically evaluated from electrical characteristics of a MOSFET [2-5], but in III-V's with relatively high interface trap density, D$_{it}$, determination of the channel charge and hence the channel mobility is quite a challenging problem. In this paper, we present results of Hall mobility measurements performed on buried InGaAs quantum well (QW) channels with HfO$_2$ high-k oxide. Although the Hall mobility measurement does not exactly match the drift or effective channel mobilities, it allows for more reliable measurements of the channel transport.

EXPERIMENT

Fig. 1 shows a schematic layout of the samples used in the experiments. The structures were grown on semi-insulating (SI) Fe-doped InP(001) substrates by molecular beam epitaxy

(MBE). The structures included 400nm-thick undoped $In_{0.52}Al_{0.48}As$ buffer layer lattice-matched to the substrate. A 2nm-thick $In_{0.53}Ga_{0.47}As$ layer was grown on the InAlAs buffer prior to the 10nm-thick compressively strained $In_{0.77}Ga_{0.23}As$ QW channel. The total barrier thickness on top of the QW was varied from 0 to 7 nm, whereas the barriers thinner than 3nm consisted of $In_{0.53}Ga_{0.47}As$ and the thicker barriers included 2nm-thick $In_{0.53}Ga_{0.47}As$ layer and $In_{0.52}Al_{0.48}As$ with top 2 monolayer thick InGaAs. Following the growth of III-V layers, a 10nm-thick high-k HfO_2 gate oxide layer was grown in-situ by reactive electron beam evaporation of metallic Hf in oxygen with a pressure of 10^{-6} Torr.

Electrical properties of the structures were measured using Van der-Pauw method from 77K to room temperature (RT), Hall measurements were performed in the magnetic field of 0.5-1 T. In order to compare buried channel mobilities with different thicknesses of the top barrier layer, we have chosen the bulk modulation doping scheme below the QW channel, separated by a 5nm-thick undoped spacer layer to reduce ionized impurity scattering.

RESULTS AND DISCUSSION

Fig.2 summarizes the data on room temperature electron Hall mobility in $In_{0.77}Ga_{0.23}As$ QW channel as a function of top barrier layer thickness. Many samples with different doping concentration and rapid thermal annealing temperatures are shown. The samples demonstrating the highest mobility were annealed at 500 or 600 ^{0}C and had sheet concentrations in the range (2-4)x10^{12} cm^{-2}. The graph shows the degradation of the maximum mobility from 11400 cm^2/Vs to 1470 cm^2/Vs in the surface channel due to remote scattering associated with the presence of gate oxide.

We further can analyze the contributions of different scattering mechanisms using temperature dependence of mobility (Fig. 3). The sample with deeply buried (d=50 nm) QW channel shows $\mu \sim T^{-1.2}$ trend which is typical for InGaAs QWs in the 80-300K temperature

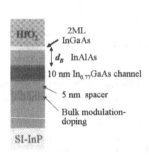
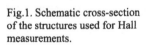
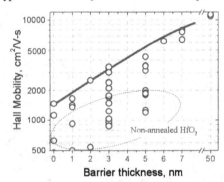

Fig.1. Schematic cross-section of the structures used for Hall measurements.

Fig.2. Room temperature Hall electron mobility in $In_{0.77}Ga_{0.23}As$ QW channel as a function of top barrier layer thickness for the samples with different carrier concentration, and annealing. Electron sheet density for the samples with the highest mobility (solid curve) is in the range of (2-4)x10^{12} cm^{-2}.

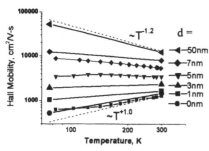

Fig.3. Temperature dependence of Hall mobility in QWs at various top barrier layer thickness.

range[6] with the main contribution from polar optical phonon scattering and some role of almost temperature-independent alloy disorder and hetero-interface roughness scattering. The samples with thinner top barriers show the reduced slope of the curves and eventually reversed slopes in the samples with d<3nm. More accurately, positive μ -T slope was observed in the samples with RT mobility below ~3000 cm^2/Vs and is close to μ ~T$^{+1.0}$ in the surface QW channels indicating the dominant role of remote Coulomb scattering (RCS) due to charges in the oxide and at the interface.

The net positive μ -T slope is a quite unexpected result that indicates considerably stronger contribution of scattering due to the HfO$_2$ interface with InGaAs than that with Si [7-9]. To test this assumption, we have prepared samples with 3nm-thick top InGaAs barrier separated from high-k oxide with 0.5-2 nm-thick amorphous Si interface passivation layers (IPL) as described elsewhere[10]. The IPL is oxidized giving rise to SiO$_x$/InGaAs interface as shown in TEM micrograph (Fig.4a). This interface is responsible for a significant increase in mobility (Fig. 5) proving that the InGaAs/HfO$_2$ interface (Fig. 4b) is a significantly stronger source for remote scattering than SiO$_x$/HfO$_2$ interface formed in the sample with a-Si passivation. This stack is qualitatively similar to the Si/HfO$_2$ interface which always contains SiO$_x$ interlayer.

As noted above, the RTA at 500-600^{0}C was typically used to improve the transport in QWs. This step is believed to reduce oxide charge due to improved microstructure and

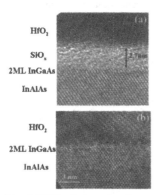

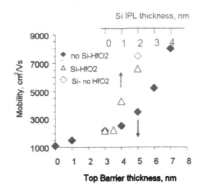

Fig. 4. TEM micrographs of interface regions of a sample: (a) with SiO$_x$ interlayer formed due to oxidation of ~2nm-thick a-Si IPL, and (b) without Si passivation.

Fig. 5. Room temperature Hall mobility in QW channel with and without SiO$_x$ interlayer and HfO$_2$ high-k oxide. Electron sheet density is close to 2x10^{12} cm^{-2}.

stoichiometry of HfO₂. However, even higher temperature is needed to reduce D_{it}, as illustrated in Fig. 6. In fact, significant charge is trapped at the interface even after 500°C, and 600°C is needed to release the trapped carriers. By accounting of the trapped carriers and associated shift of the surface potential (Fig.7), we can estimate the D_{it} present in the as-grown and 500°C-annealed samples, $\sim 6 \times 10^{12}$ cm^{-2}-eV^{-1}.

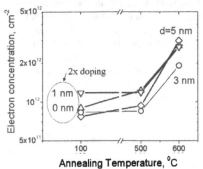

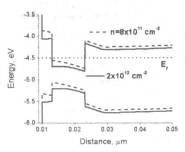

Fig. 6. Electron sheet density in the buried QW channels vs. rapid thermal annealing temperature. Modulation Si doping is about 2×10^{12} cm^{-2} for the samples with 3 and 5 nm barriers.

Fig. 7. Band diagrams of QW channel with 3 nm-thick barrier corresponding to two sheet electron concentrations. Surface potential shift is ~250 meV, corresponding to $D_{it} \sim 6 \times 10^{12}$ cm^{-2}-eV^{-1}.

The contribution of the bulk oxide charges to scattering can be assessed from Fig. 8 showing mobility dependence on HfO₂ thickness. The mobility has a clear tendency to decrease in the samples with thicker oxide. This tendency is more pronounced in non-annealed samples with higher bulk charge density. The non-gated structure revealed reduced carrier concentration (star in Fig. 8) due to high D_{it}, but channel mobility was higher than on the HfO₂-

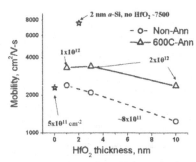

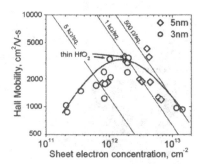

Fig. 8. Mobility vs. HfO₂ thickness for the sample with 3 nm-thick top barrier. Mobility of two samples with bare InGaAs surface and with a-Si layer (likely oxidized) on top (without HfO₂) are shown by stars.

Fig. 9. Room temperature Hall electron mobility as a function of electron density for 3nm and 5 nm-thick top barrier layers. Trend for 3nm thick barrier is shown by solid line.

gated samples. The significant improvement of mobility was observed again in the sample with SiO$_x$ interlayer as discussed above.

Fig. 9 contains room temperature Hall mobility results as a function of electron density for 3nm and 5 nm-thick top barrier layers. Here the data from samples with different doping level and annealing are shown. The trend for the highest-mobility QWs is quite obvious: the mobility has a maximum in the range $(1-4) \times 10^{12}$ cm^{-2} and reduces at lower and higher carrier densities. The low-density mobility drop is likely due to reduction of screening of the RCS by the channel electrons, and at higher densities, the electrons start occupying the high-effective-mass X-minimum.

Now we can estimate the contributions of different scattering mechanisms to the room temperature channel mobility. The samples with 50 nm-deep QW channels are not affected by the oxide interface scattering. The electron mobility is mostly limited by optical phonon scattering with some contribution of ionized impurity, hetero-interface roughness and alloy disorder scattering. The baseline mobility in the samples with bottom modulation doping is about 11500 cm^2/Vs though can be improved with top delta-doping to ~14500 cm^2/Vs or even slightly higher.

We further compare different samples with 3nm-thick InGaAs top barriers and analyze using the Mathiessen's rule how the resultant mobility is degrading:

$$\frac{1}{\mu} = \frac{1}{\mu_{phonon}} + \frac{1}{\mu_{int}} + \frac{1}{\mu_{bulk}} + \frac{1}{\mu_{Dit}} , \qquad (1)$$

where μ_{phonon} is the mobility of deep QW mostly limited by phonon scattering, μ_{int} is the scattering due to interface with oxide, μ_{bulk} is the contribution of bulk charges in the oxide, and μ_{Dit} due to scattering by the charge trapped by the interface states, the last three mechanisms acting remotely on the channel mobility.

The mobility of the samples with HfO$_2$/InGaAs interface and 3 nm-thick top barrier is limited to 3500 cm^2/Vs (Fig.9). This best mobility is observed for the annealed sample with interface trap density D$_{it}$ ~10^{12} cm^{-2}eV^{-1} and thin HfO$_2$. The scattering is therefore mostly due to Coulomb scattering by interface dipoles and fixed interface charge as well as interface roughness, and μ_{int}~ 4600-5000 cm^2/Vs . Strong (almost proportional) dependence of mobility on temperature implies a stronger contribution of the remote Coulomb scattering over the surface/interface roughness scattering, which has weaker temperature dependence.

The dependence of electron mobility on HfO$_2$ thickness (Fig. 8) allows for estimation of the contribution of the bulk oxide effects, namely oxide bulk charge and soft optical phonons. Their influence is less important, μ_{bulk} ~ 8000-15000 cm^2/Vs.

Interface states at the semiconductor surface can contribute to the remote Coulomb scattering once they are charged. However, the Fermi level tends to stay close to the conduction band at the InGaAs semiconductor surface or interface (contrary, for example, to GaAs surface) and therefore, the interface trapped charge is relatively low even at

Table 1. Mobility limits for various mechanisms at RT and 3 nm barrier (major mechanisms underlined)

Process	Limited mobility cm^2/Vs
Phonons and semi-interfaces	13,000
Interface dipoles, fixed interface charge and surface roughness	4,600-5,000
Bulk oxide charge and soft phonons (10 nm annealed HfO$_2$)	8,500-20,000
Trapped interface charge ~10^{12} cm^{-2}	>20,000

high D_{it}. The remote Coulomb scattering is studied well in high-electron mobility transistors with modulation doping, which shows insignificant mobility degradation at RT in the presence of $\sim 10^{12}$ cm^{-2} ionized impurities [11]. The effect of interface trapped charge can be also estimated from the measured mobility of the sample with a-Si passivation which has a trapped charge of the order of 10^{12} cm^{-2} but no other scattering sources associated with high-k oxide. The high value of mobility in this sample (star in Fig. 8) provides a low estimate of interface-charge-limited mobility of μ_{Dit}>20000 cm^2/Vs.

SUMMARY

In summary, we have presented Hall electron transport properties of buried QW structures influenced by remote scattering due to InGaAs/HfO$_2$ interface. Mobility degradation with reduction of the top barrier thickness is attributed primarily to remote Coulomb scattering due to charges fixed at semiconductor/oxide interface. This mechanism is responsible for a net positive mobility vs. temperature slope in the QW samples with room temperature mobility below 3000 cm^2/Vs. Introduction of SiO$_x$ interface layer formed by oxidation of thin a-Si passivation layer has been found to improve the channel mobility.

ACKNOWLEDGMENTS

The authors acknowledge financial support of INTEL Corporation and SRC Focus Center Research Program through the MSD Center.

REFERENCES

1. R. Chau, in *CS MANTECH Technical Digest*, (Chicago, IL 2008), p. 15.
2. R. J. W. Hill, D. A. J. Moran, X. Li, H. Zhou, D. Macintyre, S. Thoms, A. Asenov, P. Zurcher, K. Rajagopalan, J. Abrokwah, R. Droopad, M. Passlack, and I. G. Thayne, IEEE Electron Dev. Lett. **28**, 1080 (2007).
3. Y. Xuan, Y. Q. Wu, H. C. Lin, T. Shen, and P. D. Ye, Electron Dev. Lett. **28**, 935 (2007).
4. S. Oktyabrsky, S. Koveshnikov, V. Tokranov, M. Yakimov, R. Kambhampati, H. Bakhru, F. Zhu, J. Lee, and W. Tsai, Tech. Dig.- 65th Device Research Conference, 2007, p. 203.
5. D. H. N. Goel, S. Koveshnikov, I. Ok, S. Oktyabrsky, V. Tokranov, R. Kambhampati, M. Yakimov, Y. Sun, P. Pianetta, C.K. Gaspe, M.B. Santos, J. Lee, P. Majhi, and W. Tsai, in Tech. Dig. - Int. Electron Devices Meet. 2008, p.15.1.
6. T. Matsuoka, E. Kobayashi, K. Taniguchi, C. Hamaguchi, and S. Sasa, Jpn. J. Appl. Phys. Part 1 **29**, 2017 (1990).
7. M. A. Negara, K. Cherkaoui, P. Majhi, C. D. Young, W. Tsai, D. Bauza, G. Ghibaudo, and P. K. Hurley, Microelectron. Eng. **84**, 1874 (2007).
8. S. Barraud, L. Thevenod, M. Casse, O. Bonno, and M. Mouis, Microelectron. Eng. **84**, 2404 (2007).
9. K. Maitra, M. M. Frank, V. Narayanan, V. Misra, and E. A. Cartier, J. Appl. Phys. **102**, 114507 (2007).
10. S. Oktyabrsky, V. Tokranov, M. Yakimov, R. Moore, S. Koveshnikov, W. Tsai, F. Zhu, and J. C. Lee, Mater. Sci. Eng., B **135**, 272 (2006).
11. K. Lee, M. S. Shur, T. J. Drummond, and H. Morkoc, J. Appl. Phys. **54**, 6432 (1983).

Mater. Res. Soc. Symp. Proc. Vol. 1155 © 2009 Materials Research Society   1155-C08-03

Growth and Layer Characterization of SrTiO3 by Atomic Layer Deposition Using Sr(tBu3Cp)2 and Ti(OMe)4

Mihaela Popovici[1], S.Van Elshocht[1], N. Menou[1], J. Swerts[2], D. Pierreux[2], A. Delabie[1],

K. Opsomer[1], B. Brijs[1], G. Faelens[1], A. Franquet[1], T. Conard[1],

J. W. Maes[2], D. J. Wouters[1], J. A. Kittl[1]

[1]IMEC vzw, Kapeldreef 75, 3001-Heverlee, Belgium
[2]ASM Belgium, Kapeldreef 75, 3001-Heverlee, Belgium

ABSTRACT

Strontium titanate (STO) thin films (45-67 % Sr) were deposited by atomic layer deposition using Sr(tBu3Cp)2/Ti(OMe)4/H2O as precursors. The Sr content of the layers is well controlled by the precursor pulse ratio, as indicated by Rutherford backscattering spectroscopy (RBS). The amount of Sr and Ti deposited depends on the Sr:Ti pulse ratio and suggests the enhancement of the Ti precursor activity in the presence of Sr-OH. STO compositions that are closer to stoichiometric SrTiO3 result in denser films with correspondingly higher index of refraction. The increase (decrease) of the Sr content over (below) 50 % leads to an expansion (contraction) of the lattice parameter corresponding to cubic SrTiO3 with a perovskite structure. The dielectric constant (extracted from film thickness series) and leakage current strongly depends both on the Sr content and the crystalline state of the films.

INTRODUCTION

Strontium titanate (SrTiO3) is highly attractive as dielectric due to its high-k value for future generation complementary oxide semiconductor (CMOS) and dynamic random access memory (DRAM) applications in Metal-Insulator-Metal (MIM) capacitors. In view of the aggressive scaling of CMOS and since 3-D structures with high aspect ratio are envisaged for DRAM to achieve ultra-high storage densities, the most suitable technique is atomic layer deposition(ALD), which through its self-limiting growth mechanism ensures conformal coverage. Atomic layer deposition of STO is commonly based on the use of titanium alkoxides such as Ti(iOPr)4 [1-6] in combination with diketonates [1-8] or cycloalkenyls [9-11]. Diketonates have been mostly used, although they have the disadvantage of their low reactivity towards the most common oxidant source (H2O and O2) such that deposition temperatures of 300°C or higher are required. In this temperature range the titanium alkoxides start to decompose and the ALD window is reduced. The cyclopentadienyls have the advantage of a high volatility and reactivity towards water. Strontium-bis(tris-butylcyclopentadienyl), Sr(tBu3Cp)2, recently used in ALD [12] as the most thermally stable compound when compared with other ligands of its family [13] has recently been reported.

In this work a water-based ALD process using Sr(tBu3Cp)2 and Ti(OMe)4 precursors with high enough vapor pressure to achieve STO deposition at 250°C is reported.

EXPERIMENT

STO films were deposited in an ASM Pulsar® 3000 cross-flow ALD reactor [14] on 300 mm Si(100) wafers at 250°C using H_2O as oxidizing agent. To ensure a high enough vapor dose, the Sr and Ti precursors were heated at 180 and 160°C, respectively. The H_2O bottle was kept at 15°C. The films were grown on 300 mm Si(100)/1 nm SiO_2 and Si (100)/20nm SiO_2/10 nm ALD TiN substrates. The thickness of the films was measured with a Kla-Tencor Aset F5 equipment using a floating refractive index. Composition of the films was determined by Rutherford Backscattering Spectroscopy (RBS) in a home made endstation using a 1MeV He+ beam in a rotating random mode at a scatter angle of 170 degree. A RUMP simulation code allows the evaluation of the areal density of Sr and Ti expressed in atoms/cm^2. The as grown films having thicknesses up to 25 nm showed both thickness (standard deviation < 2.5 %) and compositional (standard deviation < 1.4 %) uniformity as determined by 49-points ellipsometry and 4-points RBS measurements, respectively. Grazing incidence X-ray diffraction (GI-XRD) was recorded with a Jordan Valley Inc. instrument for the films after a 600°C (N_2, 1min) in a Levitor (ASM) rapid thermal annealing furnace. Moreover, X-ray reflectivity measurements were recorded on the same tool, which allows the evaluation of the thickness, roughness and density of the layer stacks using the Reflectivity and Fluorescence Simulation (REFS) software package. X-ray photoelectron spectroscopy (XPS) measurements were performed on a Theta 300 system from Thermo Electron, using monochromated Al Kα photon (1486.6 eV). The energy resolution of the instrument was ~ 0.8 eV (full-width at half-maximum of the Si2 p 3/2 photoemission peak) and the binding energy calibration was made on the C 1s peak positioned at 285.0 eV. Electrical measurements in a metal-insulator-metal configuration (TiN bottom electrode-STO-Pt dots top electrode) were done on several STO compositions as deposited and annealed at 600°C.

RESULTS AND DISCUSSION

The ALD reaction cycle of the STO films is defined by the pulse sequence : $(Sr(^tBu_3Cp)_2/H_2O)_a(Ti(OMe)_4/H_2O)_b$, where a and b denote the number of deposited Sr and Ti subcycles, respectively. The precursor pulses were tuned to deposit the films in saturated conditions. We define growth per cycle (GPC) as thickness increase per one cycle. A linear growth was observed for all compositions where a:b ratios were varied between 1:1 and 7:1 (Table 1).

Table I. Composition of STO films as determined by RBS measurements at different $a:b$ pulse ratios and the corresponding GPC values (nm/cycle) obtained from ellipsometric thickness.

Pulse ratio Sr:Ti	Sr/(Sr+Ti) RBS ratio (± 0.02)	GPC (nm/cycle)
1:1	0.45	0.11
4:3	0.50	0.40
3:2	0.52	0.27
2:1	0.57	0.17
7:3	0.60	0.53
3:1	0.62	0.22
4:1	0.66	0.29
7:1	0.71	0.48

132

Stoichiometric SrTiO₃ was obtained for a pulse ratio a:b =4:3. In what follows, the compositions were simply denoted 4:3, 1:1 e.g. in accordance with the pulse ratio used. Varying the ratio a/(a+b) between 0.5-0.9, results in STO compositions with a Sr/(Sr+Ti) content between 0.45-0.71 as determined by RBS measurements.

In the Figure 1 we have plotted the RBS (atoms/nm^2) amount of metal (Sr or Ti) deposited per precursor pulse. We observed that the amount of Sr deposited only slightly decreases when the ratio Sr/(Sr+Ti) is gradually increased. At the same time, an enhanced amount of Ti is incorporated into the films, probably due to an increase of surface reactivity or more – OH sites available after Sr precursor hydrolysis. The as-deposited films are amorphous and have a refractive index between 1.83-1.85. The growth of STO films on various substrates such as SiO₂, HfO₂ and TiN has been verified measuring the amount of Sr (expressed as atoms/nm^2) deposited during the first 60 reaction cycles. Similar behavior is observed for all studied substrates, with a slight inhibition only in the case of SiO₂ substrate for the first 10 deposition cycles (Figure 2).

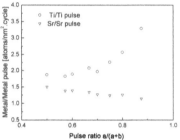

Figure 1. Amount of Sr and Ti deposited per precursor pulse versus a/(a+b) pulse ratio.

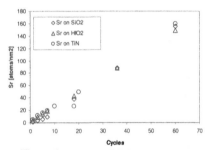

Figure 2. Amount of Sr incorporated into STO films on SiO₂, HfO₂, TiN substrates.

The sputter profile (Figure 3) for a Sr-rich STO film (2:1) indicates the presence of C below detection limit within the bulk layer. Nevertheless, the carbon is also present at the surface, (10-15%) and the surface appears to be slightly Sr-rich.

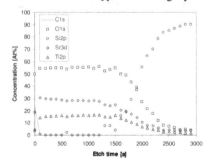

Figure 3. XPS depth profile of 2:1 STO film. The thickness of the film is ~ 18 nm.

Angle resolved XPS (data not shown) revealed that $SrCO_3$ is formed at the surface of the films ($O1s$ in $SrCO_3$ at 531.5 eV). The $C1s$ spectrum contains three components, which are typical for C surface contamination. However, the highest binding energy (BE) peak (~288.4 eV), typical for C-O bounds in carbonates is more intense as expected for typical contamination. In addition, the $O1s$ spectrum shows two components at ~ 529.5 and 531.4 eV. The low BE one corresponds to oxygen in $SrTiO_3$. The higher BE peak can also be associated with carbonate ($O1s$ in $SrCO_3$ at 531.5 eV). GI-XRD showed that the crystalline phase is cubic $SrTiO_3$ with perovskite structure, the most intense peak corresponding to the (110) diffraction line (Figure 4). The degree of crystallization as observed from the relative intensity of the peaks decreases when going towards Sr-rich STO (67 %). The diffraction lines are shifted to lower or higher angle with the incorporation of excess Sr (>50 % Sr) or Ti atoms (<50 % Sr), respectively. As result we observe variations of lattice parameter a (assuming a cubic cell for all compositions) in the range 3.896-3.976 Å (Figure 5).

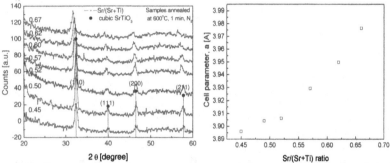

Figure 4. Grazing incidence X-ray diffractograms of STO annealed films obtained at 0.45 - 0.67 Sr/(Sr+Ti) ratio and the cubic $SrTiO_3$ pattern.

Figure 5. Lattice parameter variation as a function of Sr/(Sr+Ti) ratio assuming a cubic cell.

The crystallization anneal (600°C, 1 min) results in the increase of the refractive index from 1.83-1.85 for the as deposited amorphous STO films to values above 2.00 (Figure 6a).

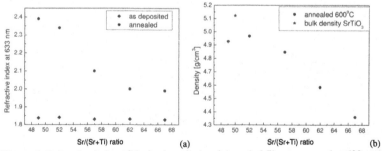

Figure 6. Refractive index of the as deposited and annealed films measured at 633 nm (a) and the density of the annealed films (b).

The maximum value is obtained for the stoichiometric $SrTiO_3$ (4:3) film which correspondingly also has the highest density (Figure 6b).

Electrical measurements

Figure 7 shows the evolution of the dielectric constant of STO films as a function of their Sr content. All films were deposited on 10 nm ALD TiN and annealed at 600°C in N_2 for 1 min. For each composition, the reported κ value was extracted from a thickness series. The dielectric properties of the STO films are highly dependent on the Sr content: the κ value being ~ 200 for the stoichiometric $SrTiO_3$ and decreasing monotonously with the amount of Sr to reach ~ 40 for a $Sr_{0.67}Ti_{0.33}O_3$ film. This trend is in good agreement with predicted values [15]. Leakage density values at +1V as a function of EOT (equivalent oxide thickness) of STO films with various compositions are presented in Figure 8. As expected, a strong leakage increase was observed with decreasing EOT. The electrical properties are dependent on the composition of the STO film. Sr-richer STO films are less leaky for comparable EOT values.

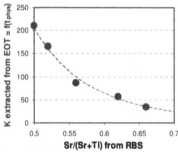

Figure 7. Variation of dielectric constant with Sr content (RBS)

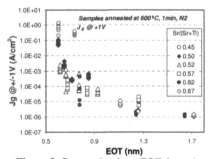

Figure 8. Current density – EOT dependence recorded at +1 V.

CONCLUSIONS

Strontium titanate (STO) thin films (45-67 % Sr) were deposited by atomic layer deposition using $Sr(^tBu_3Cp)_2/Ti(OMe)_4/H_2O$ as precursors. Compositions were varied in the range 0.45-0.67 Sr/(Sr+Ti) by changing the Sr:Ti precursor pulse ratio. An enhancement of Ti deposition is observed for the compositions richer in Sr. For all Sr:Ti ratios the films crystallized after 600°C anneal. Stoichiometric $SrTiO_3$ with perovskite structure was obtained for the 4:3 pulse ratio. An expansion of the lattice parameter of the cubic cell for Sr-richer STO is observed. Increased density and refractive index were observed for compositions closer to stoichiometric STO. Also, the highest dielectric constant was observed for stoichiometric $SrTiO_3$. The leakage current decreases for more Sr-rich STO films comparing capacitors with similar CET.

ACKNOWLEDGMENTS

Tom Blomberg (ASM Microchemistry, Finland) is kindly acknowledged for the help offered in the ALD process development.

REFERENCES

1. D.-S. Kil, J.-M. Lee, and J.-S. Roh, *Chem.Vapor Depos.* **8**, 195 (2002).
2. A. Kosola, M. Putkonen, L.-S. Johansson, and L. Niinisto, *Appl. Surf. Sci.* **211**, 102 (2003).
3. O. S. Kwon, S. K. Kim, M. Cho, C. S. Hwang, and J. Jeong, *J. Electrochem. Soc.* **152**, C229 (2005).
4. O. S. Kwon, S. W. Lee, J. H. Ham, ad C. S. Hwang, *J. Electrochem. Soc.* **154**, G127 (2007).
5. J. -H. Ahn, S.-W. Kang, J.-Y. Kim, J.-H. Kim, and J.-S. Roh, *J. Electrochem. Soc.* **155**, G185 (2008).
6. S.W.Lee, J. H. Han, O.S. Kwon, and C.S. Hwang, *J. Electrochem. Soc.* **155**, G253 (2008).
7. S.W.Lee, O.S. Kwon, J. H. Han, and C.S. Hwang, *Appl. Phys. Lett.* **92**, 222903 (2008).
8. S. Bhaskar, D. Allgeyer, and J. A. Smythe III, *Appl. Phys. Lett.* **89**, 254103 (2008).
9. M. Vehkamäki, T. Hatanpää, T. Hänninen, M. Ritala and M. Leskelä, *Electrochem. Solid-State Lett.*, **2**, 504 (1999).
10. M. Vehkamaki, T. Hanninen, M. Ritala, M. Leskela, T. Sajavaara, E. Rauhala, and J. Keinonen, *Chem.Vapor Depos.* **7**, 75 (2001).
11. A. Rahtu, T. Hanninen, and M. Ritala, *J. Phys. IV*, **11**, 923 (2001).
12. M. Vehkamaki, *PhD Thesis*, University of Helsinki, 2007.
13. T. Hatanpaa, M. Ritala, and M. Leskela, *J. Organomet. Chem.* **692**, 5256 (2007).
14. PULSAR 3000 is a trademark of ASM International.
15. S. Clima, G. Pourtois, N. Menou, M. Popovici, A. Rothschild, B. Kaczer, S. Van Elshocht, X. Wang, J. Swerts, D. Pierreux, S. De Gendt, D. Wouters, and J.A. Kittl, to be published.

Mater. Res. Soc. Symp. Proc. Vol. 1155 © 2009 Materials Research Society   1155-C09-03

Thermal Stability of GdScO₃ Dielectric Films Grown on Si and InAlN/GaN Substrates

K. Fröhlich[1], A. Vincze[2,3], E. Dobročka[1], K. Hušeková[1], K. Čičo[1], F. Uherek[2,3], R. Lupták[1], M. Ťapajna[1], D. Machajdík[1]

[1] Institute of Electrical Engineering, SAS, Dúbravská 9, 841 04 Bratislava, Slovak Republic
[2] International Laser Center, Ilkovičova 3, 841 04 Bratislava, Slovak Republic
[3] Department of Microelectronics, FEI STU, 812 19 Bratislava, Slovak Republic

ABSTRACT

We present analysis of thermal stability of thin GdScO₃ films grown on silicon and InAlN/GaN substrates. The GdScO₃ films were prepared by liquid injection metal organic chemical vapor deposition at 600 °C. The films were processed after deposition by rapid thermal annealing in nitrogen ambient at 900, 1000 and 1100 °C during 10 s. In addition, annealing of the GdScO₃ films on InAlN/GaN substrate at 700 °C during 3 hours was performed. The samples were analyzed by grazing incidence X-ray diffraction (GIXRD), X-ray reflectivity (XRR) and time-of-flight secondary ion mass spectroscopy (ToF SIMS). GIXRD confirmed that the as-deposited GdScO₃ films were amorphous. Recrystallization of the films on both substrates occurred at 1100 °C. ToF SIMS depth profile of the films annealed at 1000 °C indicated strong reaction of the GdScO₃ film with the Si substrate. For the InAlN/GaN substrate rapid thermal annealing at 900 °C induced diffusion of the In and Al atoms into the top GdScO₃ layer. Thermal treatment at 700 °C for 3 hours presents upper limit of the acceptable thermal budget for the GdScO₃/InAlN interface.

INTRODUCTION

Gadolinium scandate is considered as an alternative gate dielectric material in CMOS technology due to its κ value higher than 20. Up to now thin films of rare earth scandates films were prepared using pulsed laser deposition [1], electron beam evaporation [2], atomic layer deposition [3, 4], and metal organic chemical vapor deposition [5]. Alternatively, GdScO₃ can be used as the gate insulation in high power – high frequency high electron mobility GaN based transistors (HEMT). Integration of GaN based HEMT's with CMOS technology is a challenging objective, as it would allow new electronic applications. However, thermal stability of the dielectric film is a key property for both applications.

It was already concluded, that the advantage of the rare earth scandates thin films over other high-κ materials is, that they preserve their amorphous structure at temperatures up to 1000 °C. However, more detailed studies have revealed limited thermal stability of dysprosium scandate on Si substrate [5, 6]. It has been observed, that dysprosium scandate thin films form silicate under high temperature anneal. More experimental data is required to understand rare earth scandates thin film behavior at elevated temperatures. In our contribution we analyze properties of thin GdScO₃ films on two different substrates (Si and InAlN/GaN) submitted to high temperature post-deposition processing.

EXPERIMENT

The GdScO₃ films were deposited at 600 °C using Gd(thd)₃ and Sc(thd)₃ precursors dissolved in toluene in a concentration of 0.02 M. The reaction atmosphere was composed of O_2 (170 sccm flow rate) and Ar (21 sccm flow rate) with a total pressure of 200 Pa. Injection frequency was 0.33 Hz. Opening time of the micro valve was 3 ms, resulting in the droplet mass of 5.7 mg. Employing these parameters the film growth rate was adjusted to 0.8 nm/min. Si (100) oriented wafers and InAlN/GaN heterostructure on sapphire were used as substrates. The InAlN/GaN/sapphire substrates were composed of 16 nm thin InAlN grown on 2.5 μm thick GaN buffer. The heterostructures were grown by MOCVD by AIXTRON A.G.

Thermal stability of the GdScO₃ films was examined by rapid thermal annealing (RTA) in N_2 atmosphere. The samples were heated up using a ramp rate of 100 °C/s to 900, 1000, and 1100 °C with a dwell time of 10 s. To verify possible post-deposition processing, the GdScO₃ films grown on InAlN/GaN were annealed also at 700 °C for 3 hours.

To study recrystallization and interfacial reactions, annealed samples were analyzed by grazing incidence X-ray diffraction (GIXRD), X-ray reflectivity (XRR) and time-of-flight secondary ion mass spectroscopy (ToF SIMS). GIXRD and XRR analysis were performed on Bruker AXS–D8 DISCOVER equipment using Cu Kα radiation.

The time-of-flight based SIMS instrument (Ion-TOF) with high energy Bi^+ primary source was employed for element depth profiling. For the depth profiling high energy pulsed primary source (25 keV) was combined with low energy sputter guns at 500 eV Cs^+ gun in 45° to sample surface. Sputtering ion beam was rastered over 300×300 μm² area while the primary beam scanned through 80×80 μm² in the centre of the sputtered area. Because of slightly charging of the samples electron flooding was used.

RESULTS AND DISCUSSION

Figure 1a and Figure 2a display GIXRD and XRR of GdScO₃ films grown on Si substrates as-deposited and submitted to the rapid thermal annealing from 900 to 1100 °C. As-deposited films were amorphous with broad peak at around 30 °, typical for weakly crystalline to amorphous phase. Simulation of the XRR pattern of the as-deposited film resulted in determination of the thickness equal to 16 nm. Annealing of the film at 900 °C improved XRR pattern without any changes in GIXRD. Thickness of the film extracted from the XRR pattern simulation did not change. After processing of the films at 1000 °C the broad diffraction peak located at 30 ° has disappeared. Furthermore, shorter period of the XRR pattern modulation indicates increase of the film thickness. Simulation of the XRR pattern gave thickness of 25 nm. Recrystallization of the film appeared after rapid thermal annealing at 1100 °C. The XRR pattern has also changed substantially: modulation at higher angles disappeared.

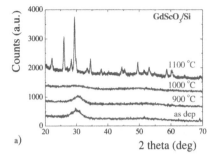
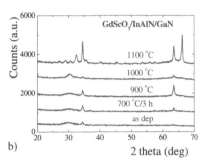

a) 2 theta (deg)   b) 2 theta (deg)

Figure 1. Grazing incidence X-ray diffraction pattern of GdScO₃ films on a) Si and b) InAlN/GaN substrates annealed at various temperatures.

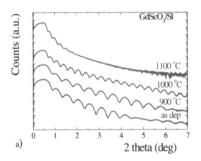
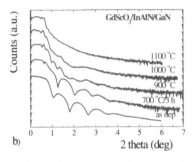

a) 2 theta (deg)   b) 2 theta (deg)

Figure 2. X-ray reflectivity pattern of GdScO₃ films on a) Si and b) InAlN/GaN substrates annealed at various temperatures.

X-ray diffraction and XRR patterns of the GdScO₃ films deposited on InAlN/GaN substrate are shown in Figure 1b and Figure 2b, respectively. Origin of two GIXRD peaks tlocated at 2 theta = 34.5 ° and 63.4 ° is not well understood. However, they are clearly connected with the diffraction from the InAlN/GaN substrate. Besides these peaks the GIXRD pattern corresponds to weakly crystalline to amorphous film. The XRR pattern of the as-deposited sample is less modulated due to rougher surface of the InAlN/GaN substrate. The XRR pattern of the GdScO₃ film processed at 700 °C/3 hours is very similar to that of the as-deposited sample. Important changes are visible on the XRR patterns after RTA above 900 °C. Annealing of the structure at 1100 °C induced recrystallization of the GdScO₃ film with a complete vanishing of the XRR pattern modulation.

ToF SIMS depth profiles of the GdScO₃ sample correspond very well to the GIXRD and XRR results. Figure 3 shows the depth profiles for the as-deposited GdScO₃/Si film. The depth profile indicates quite well defined interfaces between the GdScO₃ film and Si substrate. We note, that there is a GdSiO interface layer present already in as the as-deposited sample. The signal coming from ScSiO ions (not shown) is almost identical to that of GdSiO. Upon

139

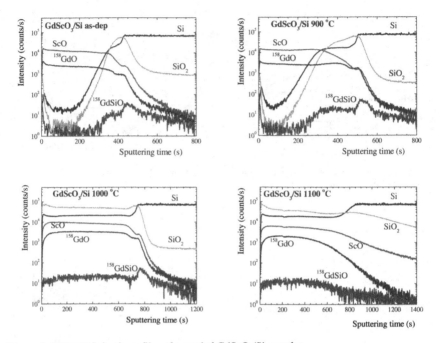

Figure 3. ToF SIMS depth profiles of annealed GdScO$_3$/Si samples.

annealing at 900 °C the interface GdSiO layer becomes broader. Further broadening of the GdSiO, SiO$_2$ interlayer and penetration of Si atoms through the whole GdScO$_3$ layer can be observed after annealing at 1000 °C. It means, that GdScO$_3$ layer transforms to GdSc-silicate upon 1000 °C annealing. The GdSc-silicate formation observed using ToF SIMS is corroborated by the results of XRR, where the modification of the signal modulation indicated apparent increase of the film thickness. RTA at 1100 °C resulted in complete intermixing of the atoms at the GdScO$_3$/Si interface.

The SIMS profiles evolution of the GdScO$_3$/InAlN sample upon annealing is shown in the Figure 4. The depth profile of the as-deposited sample presents reference behavior of the structure with well defined interfaces. The sample processed at 700 °C for 3 hours preserves its structure; even though the signal due to the AlN and In ions at the surface has increased. RTA at 900 °C induces diffusion of the AlN an In ions through the GdScO$_3$ to the top surface. AlN and In ions signal at the surface is even more pronounced after 1000 °C treatment. Similarly as in the case of GdScO$_3$/Si, RTA at 1100 °C gave rise to complete intermixing of the atoms at the GdScO$_3$/InAlN interface (not shown).

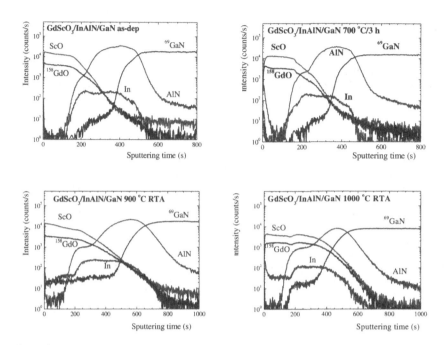

Figure 4. ToF SIMS depth profiles of annealed GdScO₃/InAlN/GaN samples.

CONCLUSIONS

In conclusions, analysis of the thermal stability of the GdScO₃/Si and GdScO₃/InAlN interfaces using GIXRD and XRR is well corroborated by depth profiles evolution observed using ToF SIMS. GdSc-silicate interlayer observed in the as-deposited GdScO₃/Si films is getting broader upon annealing. During 1000 °C RTA the GdScO₃ film completely transforms to GdSc-silicate, stabilizing thus its amorphous character. RTA at 1100 °C induces pronounced interdiffusion at the GdScO₃/Si interface.

GdScO₃/InAlN interface appeared to be less stable upon the thermal treatment. During the processing at 700 °C for 3 hours the interface is well preserved, even though diffusion of the In and AlN ions was already detected. These conditions present probably upper limit of the acceptable thermal budget for the GdScO₃/InAlN interface. Rapid thermal annealing at 900 °C gave rise to intense diffusion of the In and AlN ions into the top GdScO₃ layer. Thermal treatment at 1000 °C resulted in pronounced intermixing of the GdScO₃/InAlN interface.

ACKNOWLEDGMENTS

This work was supported by the FP7 project No. 214610 MORGaN and VEGA projects (1/0787/09 and 2/0031/08). A.V. and F.U. would like to acknowledge the support from the project CENAMOST (VVCE 0049-07). Providing of the InAlN/GaN substrate by the Aixtron A.G. is greatly acknowledged.

REFERENCES

1. C. Zhao, T. Witters, B. Brijs, H. Bender, O. Richard, M. Caymax, T. Heeg, J. Schubert, V. V. Afanasev, A. Stesmans and D. G. Schlom, *Appl. Phys. Lett.* **86** 132903 (2005).
2. M. Wagner, T. Heeg, J. Schubert, St. Lenk, S. Mantl, C. Zhao, M. Caymax and S. De Gendt, *Appl. Phys. Lett.* **88** 172901 (2006).
3. K. H. Kim, D. B. Farmer, J.-S. M. Lehn, P. V. Rao and R. G. Gordon, *Appl. Phys. Lett.* **89** 133512 (2006).
4. P. Myllymäki, M. Roeckerath, M. Putkonen, S. Lenk, J. Schubert, L. Niinistö and S. Mantl, *Appl. Phys. A* **88** 633 (2007).
5. R. Thomas, P. Ehrhart, M. Luysberg, M. Boese, R. Waser, M. Roeckerath, E. Rije, J. Schubert, S. Van Elshocht and M. Caymax, *Appl. Phys. Lett.* **89** 232902 (2006).
6. C. Adelmann, S. Van Elshocht, A. Franquet, T. Conard, O. Richard, H. Bender, P. Lehnen and S. De Gendt, *Appl. Phys. Lett.* **92** (2008).

Mater. Res. Soc. Symp. Proc. Vol. 1155 © 2009 Materials Research Society   1155-C09-15

Weichao Wang[1], Ka Xiong[1], Geunsik Lee[2], Min Huang[2], Robert M. Wallace[1,2] and Kyeongjae Cho[1,2,*]

[1] Department of Materials Science & Engineering and [2] Department of Physics,
The University of Texas at Dallas, Richardson, TX 75080
*kjcho@utdallas.edu

ABSTRACT

We investigated the HfO_2:GaAs interface electronic structure and interface passivation by first principles calculations. The HfO_2:GaAs interface of HfO_2 terminated with four O atoms and GaAs terminated two Ga atoms is found to be the most energetically favorable. It is found that the interface states mainly arise from the interfacial charge mismatch, more specifically from the electron loss of interfacial As. Si or Ge as an interfacial passivating layer helps to maintain the charge of interfacial As and hence reduce the interface states.

INTRODUCTION

Gaas has been of great interest in high performance channel metal-oxide-semiconductor field effect transistors (MOSFETs) because it has higher mobility and a higher breakdown field compared to Si-based devices. However, a major obstacle for integrating GaAs into MOS devices is the poor interface quality between the channel and the gate oxide. To improve the interface properties, there has been intensive research on understanding the origin of the interface states and possible ways to passivate them[1-4].

It is known that the GaAs native oxide[1] generates a relatively high density of interfacial states inducing Fermi level pinning[5]. Besides native oxides, extrinsic oxides such as SiO_2[2] may also contribute to SiO/GaAs interface states due to the partially filled dangling Si bonds. Currently, device scaling requires a high dielectric constant (k) gate dielectric oxide to replace conventional SiO_2. HfO_2 is one of the alternatives thanks to its high k value, superior thermal stability, and low bulk trap density. However, growing HfO_2 directly on GaAs is not expected to form a high quality interface due to the formation of GaAs native oxides at the interface. To eliminate the native oxide, Ok $et\,al$[3,4] recently reported the passivation of the GaAs:HfO_2 interface by Si and Ge. SiO_x and GeO-like compounds were formed to avoid the native oxide aggregation, thus significantly improving the equivalent oxide thickness and leakage current of the GaAs:HfO_2 interface.

Nevertheless, a precise atomic scale analysis of the origin of the interface states and the passivation mechanism is still missing as compared to the HfO_2:Si interface[6]. In this paper, we focus on the origin of intrinsic interface states and discuss the interface passivation mechanism of GaAs:HfO_2 interface by Si and Ge from a theoretical point of view.

COMPUTATATONAL METHODS

Our calculations are based on the density functional theory (DFT) method with the PW91 version of the generalized gradient approximation (GGA), as implemented in a plane-wave basis code VASP[7-9]. The pseudopotential is described by projector-augmented-wave (PAW) method[10-11]. An energy cutoff of 400 eV and a 8×8×1 k-point mesh were used. Interfaces are modeled by a superlattice containing two interfaces and no vacuum. To reduce the strong quantum size effect[12] of GaAs, the superlattice contains 7 units of HfO_2 and 21 units of GaAs.

RESULTS

We considered four HfO_2:GaAs interfaces: the interface formed by (1) O-Ga bonds (denoted as O/Ga), (2) O-As bonds (O/As), (3) Hf-Ga bonds (Hf/Ga), and (4) Hf-As bonds (Hf/As). Table I summarizes the calculated formation energies of these interfaces. The formation energy[17] is given by $E_{Form}= E_{tot(HfO2-GaAs)} - E_{HfO2} - E_{GaAs}$, where $E_{tot(HfO2-GaAs)}$ is the total energy of the relaxed system; E_{HfO2} (E_{GaAs}) is the total energy of the HfO_2(GaAs) region, obtained from a supercell calculation of the originally relaxed HfO_2-GaAs system by pulling out GaAs(HfO_2) layers and considering the energy of the structure without atomic relaxation. Our calculation shows O/Ga interface formation energy is 0.233 eV, 0.171 eV and 0.089 eV lower than As/O, Hf/As and Hf/Ga, respectively. Consequently, we will focus the analysis on the O/Ga interface.

Table I. Formation energies of four different interfaces: O/Ga, O/As, Hf/Ga and Hf/As.

	O/Ga	O/As	Hf/Ga	Hf/As
$E_{Form}(eV/Å^2)$	-0.736	-0.503	-0.648	-0.565

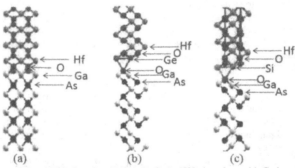

FIG.1 (a) Side view of GaAs:HfO_2 interface(O/Ga) (b) GaAs:HfO_2 interface with Ge interfacial passivation layer (IPL) (c) GaAs:HfO_2 interface with Si IPL. As is shown in black, Ga in white, Ge in pink, O in red, Si in green, and Hf in light blue.

The relaxed structure of O/Ga interface is shown in Fig. 1(a). Each interfacial Ga atom is bonded to two interfacial O atoms with bond lengths of 2.07 Å and pushes the remaining two O interfacial atoms up by 0.76 Å, similar to the relaxed structure of the HfO_2:Si interface.[5] Fig. 2 shows the calculated band structure of this interface. The band gap of bulk GaAs can be derived from the projected bulk GaAs bands (shown as black dots), which is of 1.09 eV, compared to the experimental value of 1.42 eV. It is clearly shown that this interface is metallic with 6 interfacial states in the gap (labeled as 1-6). The corresponding charge densities (Fig.3) at the Gamma point

144

of the Brillouin zone reveal that the states 1-4 are related to partially filled As bonding states, while 5 and 6 correspond to the contributions of interfacial Ga and O. Charge density plots show that the interface states are mainly contributed by the interfacial As atoms. A Bader charge analysis[14] shows that the interfacial As, Ga and O have 0.34, 0.65 and 0.05 electrons less than their corresponding bulk atoms, respectively. Band structure (Fig. 2) and charge density plots (Fig. 3) show interfacial states are mainly due to the interfacial As rather than Ga and O. Thus we can conclude that interfacial As atoms are much more sensitive to the surrounding chemical environment than interfacial Ga atoms. This finding is similar to what was found in our previous study of oxidized GaAs surfaces[13]. The origin of the interface states can then be explained in terms of local charge transfer. When GaAs is in contact with HfO_2, there exists a competition for the charge transfer due to different electron affinities of interfacial O and As. Since O has a larger electronegativity than As, interfacial As does not maintain as many valence electrons in Ga-As bond as its bulk atoms. As a result Ga-O bond is saturated, but Ga-As bond is partial occupied leading to gap state formation.

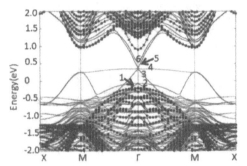

FIG.2 Two dimensional band structure for ideal HfO_2:GaAs interface configurations (O/Ga interface). The dotted region indicates the projected GaAs bulk band. Interfacial gap states are labeled as 1-6.

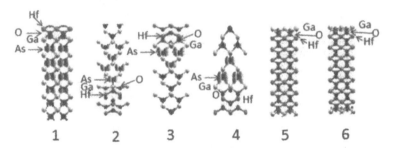

FIG. 3 Charge density distribution for six interfacial states at Γ point. Charge density contour is shown in yellow. As is shown in black, Ga in white, O in red, and Hf in light blue. Each charge density plot number corresponds to the corresponding interfacial gap state number shown in Fig. 2, respectively.

For the purpose of interface passivation, various elements are proposed to reduce interfacial gap states. Ge was found to be a promising element because Ge bonds to interfacial O atoms and prohibits the native oxide accumulation. Hinkle *et.al*[15] found that the native oxide, i.e., Ga_2O is not removed easily from the GaAs surface. To mimic this experimental condition, Ga_2O-like oxide is present in the Ge passivating interface as shown in Fig. 1(b). One O atom is bonded to two Ga atoms at the interface and the corresponding Ga-O bond length is 1.85 Å. Ge prefers to stay close to the HfO_2 surface and forms GeO-like compound. Fig. 4 shows the two dimensional band structure of the Ge passivated interface. Bands 1 and 2 are located within the GaAs bulk band gap. These two gap states could lead to p-type pinning as they lie close to GaAs valence band edge. At the Γ point, the state 2 is related to partially charged interfacial As atom (Fig. 5(b)). From the intrinsic O/Ga interface study, it is shown that the interface states originate from the interfacial As. With a Ge IPL, the interfacial As has 0.03 electrons less than its bulk atoms, compared to 0.34 electrons loss in the case of O/Ga interface. This tiny remaining charge imbalance of the interfacial As may still induce interfacial states at the Fermi level. This finding further confirms that As is very sensitive to the chemical environment. Interestingly, band 1 corresponds to interfacial As contribution except at the Γ point, where the interfacial band falls into the bulk region (Fig. 5(a)). From this analysis, we find that Ge can partially passivate the interface states. However, the interfacial charge mismatch still exists so that the interface states cannot be passivated completely. This finding is consistent with recent reports[16].

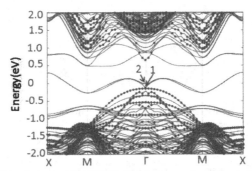

FIG.4 Two dimensional band structure for HfO_2:GaAs interface passivated by an inserting Ge layer. The dotted region indicates the projected GaAs bulk bands. Interfacial bands are labeled as 1 and 2.

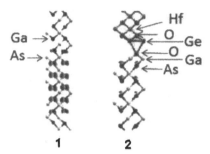

Fig. 5 Charge density distribution of two interfacial bands at Γ point. Charge density contour is shown in yellow. As is shown in black, Ga in white, Ge in pink, O in red, and Hf in light blue. Each charge density plot number corresponds to the corresponding interfacial band number shown in Fig. 4, respectively.

Similarly, a Si IPL was considered to passivate the HfO_2:GaAs interface. As shown in Fig. 6, this specific interface configuration generates four interfacial bands (labeled as 1-4), compared to two interfacial bands for the Ge IPL case. These four bands lie close to the GaAs valence band edge, and would lead to p-type pinning, similar to the Ge IPL case. Charge density plots (Fig. 7) show bands 3 and 4 are introduced by interfacial Si. Band 1 and 2 are due to the tiny charge unbalance of interfacial As. This finding is consistent with previous reports by Kummel and coworkers[2] on the GaAs surface.

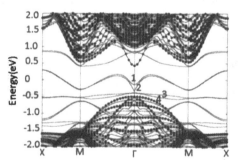

FIG.6 Two dimensional band structure for HfO_2:GaAs interface passivated by an inserting Si layer. The dotted region indicates the projected GaAs bulk bands. Interfacial bands are labeled as 1-4.

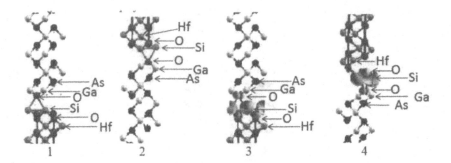

FIG.7 Charge density distribution for four interfacial bands at the Γ point. Charge density contour is shown in yellow. As is shown in black, Ga in white, Si in green, O in red, and Hf in light blue. Each charge density plot number corresponds to the corresponding interfacial band number shown in Fig. 6, respectively.

CONCLUSIONS

In summary, we have investigated the origin of interface states and interface passivation of HfO_2:GaAs by first principles calculations. The interface states arise from the interfacial charge mismatch, mainly due to the partially filled bonding state of interfacial As. Si and Ge IPL could reduce charge unbalanced of interfacial As. Therefore, they help to reduce interface states, but they do not completely passivate the gaps states due to the existence of interfacial charge mismatch. Furthermore, the IPL inhibits the presence of native oxides at the interface, thus avoiding the native oxide induced Fermi level pinning.

ACKNOWLEDGMENTS

This research is supported by the FUSION/COSAR project and the FCRP Center on Materials, Structures, and Devices. We thank the III-V materials research groups at UTD for helpful discussions, in particular Prof. Eric Vogel and Dr. Christopher Hinkle. Calculation was done on the TEXAS ADVANCED COMPUTER CENTER (TACC).

REFERENCES

[1] C.L. Hinkle, A.M. Sonnet, M. Milojevic, F.S. Aguirre-Tostado, H.C. Kim, J.Kim, R.M. Wallace, and E. M. Vogel, Appl. Phys. Lett., **93**, 113506(2008).
[2] D.L. Winn, M. J. Hale, T.J. Grassman, A. C. Kummel, R. Droopad and M. Passlack. J. Chem. Phys., **126**, 084703(2007)
[3] I. Ok, H. Kim, M. Zhang, F. Zhu, S. Park, J. Yum, H. Zhao, and J. C. Lee, Appl. Phys. Lett. **91**, 132104 (2007)
[4] H.S. Kim, I. Ok, M. Zhang, C. Choi, T. Lee, F. zhu, G. Thareja, L. Yu, and J. C. Lee, Appl. Phys. Lett., **88**, 252906(2006).
[5] J. Tersoff, Phys. Rev. Lett. 52, 465 (1984).
[6] P. W. Peacock, K. Xiong, K. Tse, and J. Robertson, Phys. Rev. B **73**,075328 (2006).

[7] G. Kresse and J. Furthmüller, Comput. Mater. Sci. **6**, 15 (1996).
[8] G. Kresse and J. Furthmüller, Phys. Rev. B **54**, 11169 (1996).
[9] G. Kresse and J. Hafner, Phys. Rev. B **47**, 558 (1993).
[10] P. E. Blochl, Phys. Rev. B 50, 17953(1994).
[11] G. Kresse and J. Hafner, J. Phys.: Condens. Matter **6**, 8245(1994).
[12] J. Jiang, B. Gao, T.-T. Han, and Y. Fu, Appl. Phys. Lett., **94**, 092110(2009).
[13] W. Wang, G. Lee, M. Huang, R.M. Wallace, and KJ Cho, 5th international Symposium on Advanced Gate State Technology, **P5**, session 4, Austin, 2008.
[14] G. Henkelman, A. Arnaldsson, and H. Jónsson, Comput. Mater. Sci. **36**, 254 (2006).
[15] C.L. Hinkle, M. Milojevic, E.M. Vogel, R.M. Wallace, Microelectron. Eng. (2009), doi:10.1016/j.mee.2009.03.030.
[16] H.S. Kim, I. Ok, F. zhu, M. Zhang, S. Park, J. yum, H. Zhao, P. Majhi, D. I. Garcia-Gutierrez, J. Goel, W. Tsai, C. K. Gaspe, M. B. Santos, and Jack C. Lee, Appl. Phys. Lett., **93**, 132902(2008).
[17] B. Magyari-kope, S. Park, L. Colombo, Y. Nishi, and Kyeongjae Cho, J. Appl. Phys. **105**, 013711(2009).

Mater. Res. Soc. Symp. Proc. Vol. 1155 © 2009 Materials Research Society 1155-C10-06

Interface and Electrical Properties of Atomic-Layer-Deposited HfAlO Gate Dielectric for N-Channel GaAs MOSFETs

Rahul Suri[1], Daniel J. Lichtenwalner[2] and Veena Misra[1]
[1]Electrical and Computer Engineering, North Carolina State University, Raleigh, NC, 27695
[2]Materials Science and Engineering, North Carolina State University, Raleigh, NC, 27695

ABSTRACT

The interface and electrical properties of HfAlO dielectric formed by atomic layer deposition (ALD) on sulfur-passivated GaAs were investigated. X-ray photoelectron spectroscopy (XPS) revealed the absence of arsenic oxides at the HfAlO/GaAs interface after dielectric growth and post-deposition annealing at 500 °C. A minimal increase in the amount of gallium oxides at the interface was detected between the as-deposited and annealed conditions highlighting the effectiveness of HfAlO in suppressing gallium oxide formation. An equivalent oxide thickness (EOT) of ~ 2 nm has been achieved with a gate leakage current density of less than 10^{-4} A/cm^2. These results testify a good dielectric interface with minimal interfacial oxides and open up potential for further investigation of HfAlO/GaAs gate stack properties to determine its viability for n-channel MOSFETs.

INTRODUCTION

Recently, III-V semiconductor materials such as GaAs, because of their high electron mobility, have gained significant attention as alternative channel materials for scaling of n-channel metal-oxide-semiconductor field effect transistors (MOSFETs). In order to achieve high performance (high speed and low power) GaAs MOSFETs at 22 nm technology node and beyond, a high quality interface between the high-k gate dielectric and GaAs substrate is imperative. Significant research efforts have focused on surface passivation of GaAs as well as finding a high quality and thermodynamically stable high-k dielectric on GaAs. HfO$_2$, by itself, on GaAs yields poor electrical characteristics because of native oxide growth during dielectric deposition and annealing [1]. The formation of native oxides (Ga and As oxides) renders the interface with a large interface state density which causes Fermi level pinning and hampers the device electrical performance. The approach of using an interface passivation (IPL) layer prior to HfO$_2$ deposition has yielded convincing results. Relatively low EOT and low gate leakage have been achieved by employing an IPL such Si [2,3], Ge [4] and Ge$_x$N$_y$ [5]. However, in order to achieve the lowest possible EOT gate stacks, and avoid issues related to having a lower mobility semiconductor IPL, it is of practical importance to investigate and establish a high-k dielectric on GaAs that yields a good quality interface without requiring an additional IPL. This necessitates the need of a surface passivation technique that effectively removes the native oxides prior to dielectric deposition and a high-k dielectric that retards its formation during dielectric growth and post-annealing. Suppression of native oxide growth at the interface is one key requirement towards obtaining high performance devices, enabling interface state density reduction and Fermi level unpinning.

In this investigation, we examine the interface and electrical properties of HfAlO dielectric formed by atomic layer deposition (ALD) on sulfur-passivated (S-passivated) p-GaAs. Ex-situ x-ray photoelectron spectroscopy (XPS) was employed to examine the changes in the interface chemical bonding caused by ALD HfAlO growth and subsequent post-deposition annealing (PDA). MOS capacitance-voltage (C-V) and current-voltage (I-V) techniques were used to study the electrical characteristics of the gate stack.

EXPERIMENT

MOS capacitors were fabricated on (100) p-GaAs (Zn-doped, 1×10^{18} cm^{-3}) substrates. The samples were dipped in an HCl:H$_2$O (1:5) solution for 1 minute for native oxide removal followed by a dip in (NH$_4$)$_2$S (40-48% by weight) aqueous solution for 1 minute at room temperature to obtain a sulfur-passivated surface. This cleaning scheme effectively inhibits the growth of arsenic oxides during dielectric deposition [6,7]. 4 nm of HfAlO dielectric was deposited by ALD at 200°C by alternating 3 cycles of Al$_2$O$_3$ and 3 cycles of HfO$_2$. We have found that this deposition scheme of Al$_2$O$_3$ (3 cycles)/ HfO$_2$ (3 cycles) is more effective in suppressing Ga oxide formation than Al$_2$O$_3$ (1 cycle)/ HfO$_2$ (1 cycle) [1]. The precursors used were Hf(N(CH$_3$)$_2$)$_4$ + H$_2$O for HfO$_2$ and Al(CH$_3$)$_3$ + H$_2$O for Al$_2$O$_3$. PDA was performed by rapid thermal annealing (RTA) in N$_2$ ambient. 40 nm thick TaN capped with 90 nm W was deposited by rf sputtering, and devices were patterned using reactive ion etching.

Electrical measurements were performed using HP4284A LCR meter for C-V measurements and HP4155 semiconductor parameter analyzer for I-V measurements. EOTs were extracted using the NCSU C-V program at 1 MHz taking the GaAs substrate properties and quantum mechanical correction into account. In order to investigate the interfacial chemical bonding, 3 nm of HfAlO film was analyzed by XPS using a monochromatic Al Kα x-ray source and 90° take-off angle. Peak fitting was done using the Fitt program [8]. The As 3d peak from GaAs was fitted using a doublet with a ratio of 3:2 and a separation of 0.7 eV.

DISCUSSION

XPS analysis of HfAlO on S-passivated GaAs

The chemical bonding at HfAlO/S-passivated GaAs interface was examined using XPS for HfAlO films in as-deposited state and after PDA using RTA at 500 °C for 1 minute. Fig. 1 shows As 3d spectra for the two cases. The peak at 40.9 eV corresponds to As-Ga bonding from the substrate whereas the peak at 41.6 eV is due to elemental arsenic. However, peaks due to arsenic oxides, As$_2$O$_3$ and As$_2$O$_5$, which manifest in As 3d spectra at 44.2 eV and 45.0 eV respectively are completely absent. Thus, no arsenic oxides are detected either after HfAlO deposition or after PDA at 500 °C. While elemental arsenic is detected at the HfAlO/GaAs interface in as-deposited state, its level is reduced to below XPS detection limit after PDA. The absence of arsenic oxides after ALD HfAlO deposition is in agreement with previous studies of ALD Al$_2$O$_3$ deposition on native GaAs [9] and S-passivated GaAs [10]. During ALD, a ligand exchange reaction between TMA and arsenic oxides causes the reduction of arsenic oxides to aluminum oxide [9]. Milojevic et al [11] have shown that the residual arsenic oxides after the sulfur-based clean are reduced following the first TMA precursor pulse. Thermodynamically, the formation of As$_2$O$_3$ is less favorable compared to Ga$_2$O$_3$. This is because of a lower Gibb's free energy (ΔG) per O atom of -998 kJ/mol for Ga$_2$O$_3$ than -593 kJ/mol for As$_2$O$_3$. Also, the reactions of As_2O_3 with GaS and GaAs to form Ga_2O_3 and elemental arsenic are thermodynamically favorable. During PDA, any As_2O_3 will get converted to Ga_2O_3 and elemental arsenic. Elemental arsenic has a high vapor pressure and is desorbed easily at temperatures above 350 °C [12]. This explains the suppressed level of elemental arsenic after PDA. The resulting HfAlO/GaAs interface after PDA with no arsenic oxides and no elemental arsenic is beneficial for obtaining low EOT and small hysteresis [6].

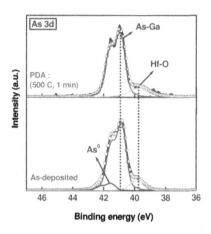

Figure 1. XPS As 3d spectra of HfAlO/S-passivated GaAs interface after dielectric growth (as-deposited) and after post-deposition annealing using RTA at (500 °C, 1min)

Fig. 2 shows the $Ga2p_{3/2}$ spectra for as-deposited and annealed HfAlO/GaAs interface. The peak at 1117.1 eV is due to the Ga-As bonding from the substrate whereas the peak at 1118.4 eV corresponds to Ga-O bonding attributed to Ga^{3+} oxidation state. The presence of sulfur at the interface is detected by virtue of a peak at 1117.8 eV corresponding to Ga-S bonding The increase in Ga-S bonding after PDA is expected to be due to S, initially loosely bonded to elemental As, bonding to Ga as As evaporates. In addition, the $Ga2p_{3/2}$ spectra is also expected to contain a contribution from Ga_2O (Ga^{+1} oxidation state) which is a suboxide (GaO_x with x<1) of Ga. However, due to the closeness of its binding energy peak position (0.55 eV from Ga-As peak) to the Ga-S peak position (0.7 eV from Ga-As peak), it is difficult to resolve a peak due to Ga_2O [11]. Hinkle et al [13] have shown that the presence of Ga_2O at the interface does not cause Fermi level pinning whereas Ga_2O_3 is detrimental to the electrical characteristics. From Fig. 2 it is inferred that between the as-deposited state and the annealed state, there is not much difference in the amount of Ga-O bonding. The percentage of Ga-O peak area relative to the total $Ga2p_{3/2}$ area is 6.5 % for as-deposited case and increases to only 9 % after PDA at 500 °C. This is in contrast to HfO_2/GaAs interface in which case a 24 % Ga-O to Ga2p percentage was observed after a similar PDA condition. Thus, HfAlO is effective in suppressing the formation of Ga oxides during dielectric growth as well as during PDA. This attribute of HfAlO distinguishes it from HfO_2 in revealing a good quality interface on GaAs.

153

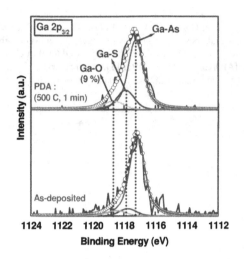

Figure 2. XPS Ga2p$_{3/2}$ spectra of HfAlO/S-passivated GaAs interface after dielectric growth (as-deposited) and after post-deposition annealing using RTA at (500 °C, 1min)

Electrical characteristics of HfAlO on S-passivated p-GaAs

Fig. 3a shows the C-V plot from a 100 um x 100 um MOS capacitor with 4 nm of HfAlO dielectric on S-passivated GaAs. The dielectric received a post-deposition anneal using RTA at 600 °C for 10s. No post-metallization annealing (PMA) was performed.

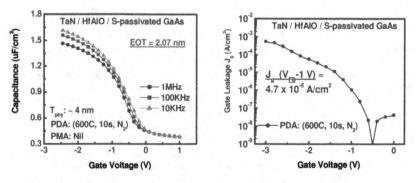

Figure 3. a) Capacitance-voltage and b) Gate leakage plots of 4 nm HfAlO/GaAs MOS capacitor. The dielectric received PDA at 600 °C for 10s. MOSCAP gate area is 1×10^{-4} cm^2

The frequency dispersion (ΔV) in depletion is defined as the gate voltage difference between the 1 MHz and 10 kHz curves at the flatband capacitance corresponding to the 1 MHz curve. The frequency dispersion in accumulation (ΔC), is the percentage difference in accumulation capacitance between the 1 MHz and 10 kHz curves. ΔV and ΔC are determined to be 0.127 V and 10% respectively. The hysteresis voltage measured between -1 V and +2.5 V at flatband capacitance corresponding to the 1 MHz curve is 0.24 V. The device EOT is extracted to be 2.07 nm. Fig. 3b shows the gate leakage current density versus the gate voltage plot which yields a leakage current density of $4.7 \times 10^{-5} A/cm^2$ at V_{fb}-1V. It is encouraging that both low EOT and low gate leakage are achieved with HfAlO dielectric without an additional IPL. The hysteresis value of 0.24 V is less than that reported for HfO_2 dielectric of same thickness with Ge IPL [4]. In addition, we observed that HfAlO/GaAs gate stack is stable up to a PDA at 800 °C with EOT of ~ 2 nm and gate leakage current density of the order of $1 \times 10^{-3} A/cm^2$.

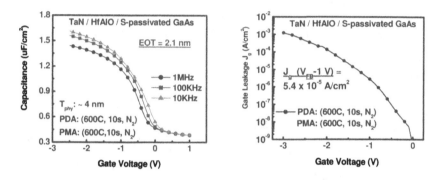

Figure 4. a) Capacitance-voltage and b) Gate leakage plots of 4 nm HfAlO/GaAs MOS capacitor after PDA and PMA both at 600 °C for 10s. MOSCAP gate area is $1 \times 10^{-4} cm^2$

Fig. 4a shows the C-V plot from the MOS capacitor with 4 nm of HfAlO dielectric after receiving a PDA at (600 °C, 10s) and a PMA at (600 °C, 10s). The device EOT is 2.1 nm. Compared to no PMA case, there is degradation observed in frequency dispersion characteristics with ΔV and ΔC increasing to 0.201 V and 12 % respectively. The gate leakage current density at V_{fb}-1V is $5.4 \times 10^{-5} A/cm^2$ which is reasonable for thin EOT. The hysteresis voltage between +1 V and -2.5 V reduced to 0.22 V after the PMA. A further optimization of PDA and PMA conditions is necessary to improve the C-V characteristics.

CONCLUSIONS

The interface and electrical properties of HfAlO dielectric on sulfur-passivated GaAs were examined using XPS and C-V and I-V measurement techniques. No arsenic oxides were detected either after dielectric deposition or PDA up to 500 °C. Only a small amount of Ga-O bonding was observed in as-deposited and annealed conditions. HfAlO effectively suppresses gallium oxide growth during deposition and PDA. With 4 nm of HfAlO, EOT of ~ 2nm with gate leakage current density of less than 1×10^{-4}A/cm^2 were achieved with small hysteresis and acceptable frequency dispersion characteristics. By optimizing the post-deposition and post-metallization conditions, the dielectric properties can be improved further. HfAlO on sulfur-passivated p-GaAs yields a good interface without an additional interface passivation layer. Further investigation is needed to determine the suitability for n-channel GaAs MOSFETs.

ACKNOWLEDGMENTS

The authors would like to thank Applied Materials, National Science Foundation and NCSU Nanofabrication Facility for their support in this work.

REFERENCES

1. R. Suri, B. Lee, D. J. Lichtenwalner, N. Biswas, and V. Misra, APL 93, 193504(2008).
2. S. Koveshnikov, W. Tsai, I. Ok, J. C. Lee, V. Torkanov, M. Yakimov, and S. Oktyabrsky, Appl. Phys. Lett. **88**, 22106 (2006).
3. R. Kambhampati, S. Koveshnikov, V. Tokranov, M. Yakinov, R. Moore, W. Tsai, and S. Oktyabrsky, ECS Transactions, 11 (4), 431-439 (2007).
4. Hyoung-Sub Kim, Injo Ok, Manhong Zhang, F. Zhu, S. Park, J. Yum, Han Zhao, and Jack C. Lee, Appl. Phys. Lett. **91**, 042904 (2007).
5. H. Zhao, H.-S. Kim, F. Zhu, M. Zhang, I. OK, S. Park, J. H. Yum, and J. C. Lee, Appl. Phys. Lett. **91**, 172101 (2007).
6. R. Suri, D. J. Lichtenwalner, and V. Misra, APL 92, 243506 (2008).
7. D. J. Lichtenwalner, R. Suri, and V. Misra, Mater. Res. Soc. Symp. Proc. Vol. 1073, (2008).
8. Hyun-Jo Kim, Hyeong-Do Kim, 'Fitt' program for XPS curve analysis, http://escalab.snu.ac.kr/~berd/Fitt/fitt.html, accessed July 2008.
9. C. L. Hinkle, A. M. Sonnet, E. M. Vogel, S. McDonnell, G. J. Hughes, M. Milojevic, B. Lee, F. S. Aguirre-Tostado, K. J. Choi, H. C. Kim, J. Kim, and R. M. Wallace, Appl. Phys. Lett. **92**, 071901(2008)
10. D. Shahrjerdi, E. Tutuc, and S. K. Banerjee, Appl. Phys. Lett. **91**, 063501 (2007)
11. M. Milojevic, C. L. Hinkle, F. S. Aguirre-Tostado, H. C. Kim, E. M. Vogel, J. Kim, and R. M. Wallace, Appl. Phys. Lett. **93**, 252905 (2008)
12. I. Karpov et al, J. Vac. Sci. & Tech. B: Microelectronics and Nanometer Structures,13, 1933 (1995)
13. C. L. Hinkle, M. Milojevic, B. Brennan, A. M. Sonnet, F. S. Aguirre-Tostado, G. J. Hughes, E. M. Vogel, and R. M. Wallace, Appl. Phys. Lett. **94**, 162101 (2009)

Mater. Res. Soc. Symp. Proc. Vol. 1155 © 2009 Materials Research Society   1155-C06-02

Thermodynamics and Kinetics for Suppression of GeO Desorption by High Pressure Oxidation of Ge

K. Nagashio[1,2], C. H. Lee[1], T. Nishimura[1,2], K. Kita[1,2] and A. Toriumi[1,2]

[1]The University of Tokyo, 7-3-1 Hongo, Bunkyo-ku, Tokyo 113-8656, Japan

[2]JST-CREST, 7-3-1 Hongo, Bunkyo-ku, Tokyo 113-8656, Japan

ABSTRACT

We analyze a main scheme for the suppression of GeO desorption by the high pressure oxidation which drastically improve the electrical quality of Ge/GeO_2 capacitors. The inherent driving force for GeO to form at the Ge/GeO_2 interface and to diffuse toward the GeO_2 surface was realized by the concentration gradient in the GeO_2 film, which was obtained from the thermodynamic calculation. Kinetic consideration based on the comparison with Si/SiO_2 stacks suggests that GeO desorption at the GeO_2 surface is the rate-limiting process under passive oxidation conditions. When O_2 pressure is increased by high pressure oxidation, the vapor pressure of GeO at the GeO_2 surface is reduced, restricting GeO desorption at the GeO_2 surface.

1. INTRODUCTION

The introduction of high-k dielectrics into the Si-CMOS process makes Ge very attractive for applications in field effect transistors, since Ge possesses an intrinsically higher carrier mobility than that of Si and the historical deterioration in quality of the Ge/GeO_2 interface due to GeO desorption [1,2] can be overcome by replacing GeO_2 to high-k dielectrics [3]. Although several techniques (e.g., surface nitridation [4] and passivation with several Si layers [5]) have been proposed for Ge/high-k stacks, the best intrinsic quality of Ge/GeO_2 interface should be realized since the interface layer composed of GeO_x between Ge and high-k dielectrics are quite important [6]. Recently, we have shown that C-V hysteresis in Ge/GeO_2 MIS capacitors was dramatically improved from ~1V to ~0.1V by high pressure oxidation (HPO) at P_{O2} ~70 atm at 550 °C [7]. Although we suggested that GeO desorption was suppressed by high pressure oxidation, a detailed mechanism was not presented yet.

The oxidation of Si has been extensively investigated, and the volatility of SiO has been noted well [8-10]. At low oxygen pressures (i.e., ultrahigh vacuum conditions) and high temperatures, the interaction of O_2 with a clean Si surface results in a rapid flux of volatile SiO molecules away from the surface via the reaction $2Si(s) + O_2(g) = 2SiO(g)$ and the subsequent formation of a non-protective SiO_2 smoke [9,10]. This active oxidation is a promising way to produce high quality and atomically clean Si surfaces [11]. On the other hand, at high oxygen pressures and high temperatures, a continuous SiO_2 layer is formed on the Si surface via the reaction $Si(s) + O_2(g) = SiO_2(s)$, which is known as passive oxidation. The active-passive transition occurs if the oxygen partial pressure P_{O2} is equal to P_{SiO}, (e.g., $P_{O2}=P_{SiO}=10^{-5.48}$ atm at 900 °C) [9,10]. Once a Si surface is covered with a thin SiO_2 film under typical oxidation conditions, the SiO_2 film is very uniform and stable, creating extremely high electrical quality and forming the basis for Si-CMOS technology.

In the previous work [17], the electrical quality of Ge/GeO$_2$ MIS capacitors formed by thermal oxidation even under the passive condition of P_{O2} = 1 atm showed deteriorated quality due to desorption of GeO. Although Deal and Grove [12] established a well-known thermal oxidation model for Si, including the diffusion of oxidant species through the SiO$_2$ film and the chemical reaction at the Si/SiO$_2$ interface, GeO desorption must be considered in the Ge oxidation process.

In the present study, vapor pressures of volatile species, Ge(g), GeO(g) and GeO$_2$(g) are first calculated in a manner similar to Si calculations [9] based on a thermodynamic database [13]. The kinetics of both oxidation and desorption are then considered based on concentration gradients of Ge, GeO, GeO$_2$ and O$_2$ in the GeO$_2$ film constructed under the assumption that thermodynamic equilibrium is established at both the Ge/GeO$_2$ and GeO$_2$/gas interfaces [14]. Finally, we discuss the ability of HPO to suppress GeO desorption in comparison to the Si/SiO$_2$ system.

2 THERMODYNAMIC CALCULATIONS OF VAPOR PRESSURES

Figure 1 shows the O$_2$ vapor pressure - temperature diagrams for the Ge-O and Si-O systems. The circles and squares with solid lines give the equilibrium O$_2$ pressures for Ge and Si oxidation according to the following oxidation reactions;

$$Ge(s) + O_2(g) = GeO_2(s), \qquad (1)$$
$$Si(s) + O_2(g) = SiO_2(s). \qquad (2)$$

It is clear that SiO$_2$(s) is more stable than GeO$_2$(s) even in the reduced O$_2$ pressure region.

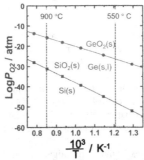

Figure 1.   O$_2$ vapor pressure - temperature diagrams for the Ge-O and Si-O systems.

Figure 2(a) shows the equilibrium vapor pressures of GeO(g), GeO$_2$(g) and Ge(g), (P_{GeO}, P_{GeO2} and P_{Ge}, respectively) as a function of P_{O2} at 550°C. This temperature is selected due to the experimental conditions proper for HPO [7]. These vapor pressures on the dotted vertical line at 550°C are presented in Fig. 1. At P_{O2}<$10^{-26.3}$ atm, GeO(g), GeO$_2$(g) and Ge(g) equilibrate with Ge(s) by:

$$Ge(s) + 1/2O_2(g) = GeO(g), \qquad (3)$$
$$Ge(s) + O_2(g) = GeO_2(g), \qquad (4)$$
$$Ge(s) = Ge(g). \qquad (5)$$

158

On the other hand, GeO(g) and GeO$_2$(g) equilibrate with GeO$_2$(s) at P_{O2}>10$^{-26.3}$ atm by:

$$GeO_2(s) = GeO(g) + 1/2O_2(g),\qquad(6)$$
$$GeO_2(s) = GeO_2(g),\qquad(7)$$
$$GeO_2(s) = Ge(g) + O_2(g).\qquad(8)$$

In above-six reactions, O$_2$(g) is not included in the reactions (5) and (7), resulting in constant vapor pressures of Ge(g) and GeO$_2$(g). Similar relations hold for Si at 900°C, as shown in Fig. 2(b). We note that the calculated temperatures for both Ge and Si are ~0.7T$_M$, typically for the dry oxidation. Figure 2(a) presents several important results: (i) P_{GeO} is the highest at the Ge/GeO$_2$ interface, (ii) P_{GeO} has a negative slope against P_{O2}, and (iii) P_{GeO} is higher than P_{Ge}.

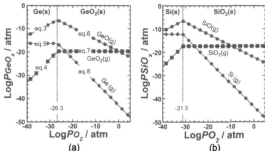

Figure 2. (a) Equilibrium vapor pressures of GeO(g), GeO$_2$(g) and Ge(g), P_{GeO}, P_{GeO2} and P_{Ge}, as a function of P_{O2} at 550°C. (b) P_{SiO}, P_{SiO2} and P_{Si} as a function of P_{O2} at 900°C.

3. KINETIC CONSIDERATION OF OXIDATION & DESORPTION

The oxidation model for Si developed by Deal and Grove [12] focuses only on the oxidation process. In case of Ge, both oxidation and desorption processes must be simultaneously considered. First, the kinetics of the oxidation and desorption processes is considered based on Wagner's theory [14], where two conditions are assumed. One is that the diffusion of atoms is the rate-limiting process (that is, thick GeO$_2$ film is assumed.). The other is that thermodynamic equilibrium is established at the Ge/GeO$_2$, and GeO$_2$/gas interfaces. Therefore, the concentrations of Ge, O$_2$, and GeO at both interfaces can be written in the range obeying Henry's law as

$$C_i = K_i P_i,\qquad(9)$$

where K$_i$ is proportional constant and species i is Ge, O$_2$ or GeO. Then, the concentration gradients are established across the GeO$_2$ film under the steady state condition. Figure 3(a) shows the concentration profiles of O$_2$, Ge, GeO and GeO$_2$ in the GeO$_2$ film.

During oxidation reactions, Ge and O$_2$ tend to migrate across the GeO$_2$ film in opposite directions and react at each interface. In case of Si, the oxidation proceeds by the inward movement of oxidant species rather than by the outward movement of silicon [12]. Similarly, only the diffusion of O$_2$ molecules was considered in Ge oxidation since the activation energy of effective oxygen diffusion in GeO$_2$ glass is reported as 1.2 eV [15] that is very similar to the case in SiO_2 glass [16]. On the other hand, for GeO desorption, GeO forms at the Ge/GeO_2 interface via the reaction

$$Ge(s) + GeO_2(s) = 2GeO \quad (10)$$

due to result (i) described in Figure 2(a), and GeO desorbs at the GeO_2 surface after diffusing through the GeO_2 film because of the negative slope (result (ii) from Figure 2(a)). This mechanism supports the recent experiments studying GeO formation at the Ge/GeO_2 interface by thermal desorption spectroscopy [17]. It should be noted that this model does not indicate that the "GeO molecule" really forms at the Ge/GeO_2 interface and diffuses. Rather, it just suggests that the driving force for a hypothetical GeO molecule exists. Moreover, GeO_2desorption can be ignored even if it takes place since it always occurs at the GeO_2 surface due to no concentration gradient in the GeO_2 film. Unlike GeO, GeO_2 desorption does not deteriorate the Ge/GeO_2 interface. Figure 3(b) summarizes the concentration profiles of effectively important species O_2 and GeO in the GeO_2 film. It is important to note that the driving force for GeO diffusion to the GeO_2 surface exists even during thermal oxidation under the passive conditions of $P_{O2} = 1$ atm.

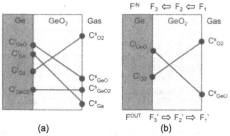

Figure 3. Concentration profiles of O_2, Ge, GeO and GeO_2 in the GeO_2 film.

For very thin oxide films, the diffusion flux of atoms is very rapid, and so the reaction rate at the interface or surface is possible to be the rate-limiting process. In this case, the equilibrium at the interface and surface should be released [18]. For general case of both thin and thick oxide films, as in the Deal and Grove model, the steady state condition is only assumed and six fluxes are considered in case of Ge, as shown in Fig. 3(b). For O_2 transport, the flux F_1 from the gas to the GeO_2 surface, the flux F_2 for the diffusion in the GeO_2 film and the flux F_3 for the reaction at the Ge/GeO_2 interface are considered. Inverse transport of GeO is similarly labeled F_1', F_2', and F_3'. Here, assuming that desorption and oxidation are independent, we only focus on the desorption process. The reaction of GeO formation is proportional to the concentration of Ge and GeO_2 atoms at the interface,

$$F_3' = k\, C_{Ge}C_{GeO2}. \quad (11)$$

The flux of GeO from the interface to the surface is given by

$$F_2' = D\frac{C_{GeO}^i - C_{GeO}^s}{x}, \quad (12)$$

where D is the diffusion coefficient for GeO and x is the GeO_2 film thickness. Finally, the flux from the surface to the gas is given by

$F_1' = h\, C^s_{GeO}.$ (13)

When the steady state condition is assumed, these three fluxes must be equal,

$$kC_{Ge}C_{GeO2} = D\frac{C^i_{GeO}}{x} = hC^s_{GeO}.$$ (14)

To understand the desorption process under the passive condition of $P_{O2} = 1$ atm, it is important to determine the rate-limiting process. The experimental determination of the rate-limiting process for desorption is difficult because the desorption process is hard to separate from the oxidation process. It is helpful to compare the Si/SiO_2 system with the Ge/GeO_2 system. Figure 4(a) compares P_{SiO} at 900 °C and P_{GeO} at 550 °C as a function of P_{O2}. The negative slope of P_{SiO} against P_{O2} indicates that there is a driving force for SiO diffusion from the Si/SiO_2 interface to the SiO_2 surface. However, unlike GeO, no SiO desorption is detected experimentally during passive oxidation. Therefore, the comparison between Si and Ge hints at the main factor affecting GeO desorption (i.e., the rate-limiting process).

F_3' for SiO is comparable to GeO since the atomic configurations at the interface are the same. The flux F_2' for SiO diffusion in the SiO_2 film is mainly governed by the SiO concentration at the interface according to eq. (12). This results in the comparable flux to GeO because P_{SiO} at the interface is almost the same as P_{GeO}. Finally, when the flux F_3' from the surface to the gas is compared in the two systems, the flux for SiO is much lower than the flux for GeO because P_{SiO} is ~ 2.5 orders of magnitude lower than P_{GeO} at the oxide film surface (at $P_{O2} = 1$ atm). Therefore, no SiO desorption is experimentally detected because of the lower P_{SiO} at the SiO_2 surface, suggesting that the rate-limiting process is desorption at the oxide surface.

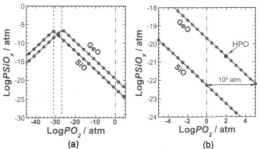

Figure 4. (a) Comparison of P_{GeO} and P_{SiO} as a function of P_{O2} at ~0.7T_M. (b) Magnified figure of (a) around P_{O2}=1 atm.

Finally the effect of HPO was considered. If HPO (for example, $P_{O2} = 100$ atom) is applied to the Ge/GeO_2 system, P_{GeO} at the GeO_2 surface is reduced in addition to the facilitation of the Ge oxidation, as shown in Fig. 4(b). This restricts GeO desorption because the flux F_1' is reduced. Moreover, P_{O2} should equal 10^5 atm to reduce P_{GeO} at the GeO_2 surface to the same level as P_{SiO} at the SiO_2 surface at P_{O2}=1 atm, as shown in Fig. 4(b). However, since the Si/SiO_2 stack is electrically stable during annealing at $P_{O2} = 10^{-3}$

atm, P_{O2} = 100 atom may be sufficient to reduce GeO desorption to levels comparable to SiO. Moreover, the behavior of GeO desorption under N_2 or UHV annealing will be different than under passive oxidation conditions. Under these conditions, P_{GeO} at the GeO_2 surface is considerably larger since remaining O_2 partial pressure determines P_{GeO}. In this case, the rate-limiting process alters GeO diffusion in the GeO_2 film [19].

4. CONCLUSIONS

To understand the mechanism for the suppression of GeO desorption by HPO, the concentration gradients of GeO from the Ge/GeO_2 interface to the GeO_2 surface were revealed based on a thermodynamic analysis of vapor pressures. Kinetics were considered by comparing the Ge/GeO_2 system with Si/SiO_2 one, suggesting that the desorption of GeO at the GeO_2 surface is the rate-limiting process under passive oxidation conditions. When the O_2 pressure is increased by the high pressure oxidation, the vapor pressure of GeO at the GeO_2 surface is reduced, restricting GeO desorption at the GeO_2 surface.

ACKNOWLEDGEMENTS

This work was partly performed in collaboration with STARC.

REFERENCES

1. J. T. Law, and P. S. Meigs, *J. Electrochem. Soc.* **104**, 154 (1957).
2. R. E. Schlier, and H. E. Farnsworth, *J. Chem. Phys.* **30**, 917 (1959).
3. C. O. Chui, S. Ramanathan, B. B. Triplett, P. C. McIntyre, and K. C. Saraswat, *Tech. Dig. IEDM,* 437 (2002).
4. A. Ritenour , S. Yu, M. L. Lee, N. Lu, W. Bai, A. Pitera, E. A. Fitzgerald, D.-L. Kwong, and D. A. Antoniadis, *Tech. Dig. IEDM,* 433 (2004).
5. W. P. Bai, N. Lu, and D.-L. Kwong, *IEEE EDL* **26**, 378 (2005).
6. Y. Kamata, *Materials Today* **11**, 30 (2008).
7. C.H. Lee, T. Nishimura, K. Nagashio, K. Kita, and A. Toriumi, *ECS Trans.*(in press).
8. C. Wagner, *J. Appl. Phys.* **29**, 1295 (1958).
9. E. A. Gulbransen and S. A. Jansson, *Oxidation of Metals* **4**, 181 (1972).
10. F. W. Smith and G. Ghidini, *J. Electrochem. Soc.* **129**, 1300 (1982).
11. A. Ishizuka and Y. Shiraki, *J. Electrochem. Soc.* **133**, 666 (1986).
12. B. E. Deal and A. S. Grove, *J. Appl. Phys.* **36**, 3770 (1965).
13. I. Barin, Thermodynamical Data of Pure Substances, PartI&II, VCH Verlags Gesellschaft, Weinheim, 1993.
14. N. Birks and G. H. Meier, Introduction to high temperature oxidation of metals, Edward Arnold, 1983, p. 42.
15. M. K. Schurman and M. Tomozawa, *J. Non-Crystal. Soids* **202**, 93 (1996).
16. J. D. Kalen, R. S. Boyce, and J. D. Cawley, *J. Am. Ceram. Soc.* **74**, 203 (1991).
17. K. Kita, S. Suzuki, H. Nomura, T. Takahashi, T. Nishimura, and A. Toriumi, *Jpn. K. Appl. Phys.* **47**, 2349 (2008).
18. K. A. Jackson, Kinetic Processes, Wiley-VCH, 2004, p.97.
19. K. Kita, C. H. Lee, T. Nishimura, K. Nagashio, and A. Toriumi, *ECS Trans.* **16**, 187 (2008).

Mater. Res. Soc. Symp. Proc. Vol. 1155 © 2009 Materials Research Society 1155-C06-07

Molecular Beam Epitaxy study of a common a-GeO$_2$ interfacial passivation layer for Ge- and GaAs-based MOS heterostructures

C. Merckling[1], J. Penaud[2], F. Bellenger[1,3], D. Kohen[1], G. Pourtois[1], G. Brammertz[1], M. Scarrozza[1], M. El-Kazzi[4], M. Houssa[3], J. Dekoster[1], M. Caymax[1], M. Meuris[1], M.M. Heyns[1,3]

[1]IMEC, Kapeldreef 75, B-3001 Leuven, Belgium
[2]Riber, 31 rue Casimir Périer, F-95873 Bezons, France
[3]K.U. Leuven, B-3001 Leuven, Belgium
[4]SUN Synchrotron, L'Orme des Merisiers, F-91196, Gif-sur-Yvette, France

ABSTRACT

Future CMOS technologies will require the use of substrate material with a very high mobility. Therefore, the combination of Ge pMOS with GaAs nMOS devices is investigated for its possible use in advanced CMOS applications. In this work, the physical, chemical and electrical properties of a-GeO$_2$ interfacial passivation layer (IPL) for n-Ge(001) and p-GaAs(001) have been investigated, using Molecular Beam Epitaxy (MBE) technique. The efficient electrical passivation of Ge/GeO$_2$ will be demonstrated, and in the case of GaAs, the use of a thin a-GeO$_2$ interlayer reduces the defects at the interface.

INTRODUCTION

Future improvements in MOSFETs performances will require high mobility (high-μ) semiconductor channels. The integration of novel materials with higher carrier mobility, to increase drive current capability, is a real challenge to overcome silicon-based CMOS. One solution is to use a germanium-based channel for pMOS combined with a III-V-based channel for nMOS. The main issues of such devices consist in obtaining low leakage current, low interface state density and high carrier mobility in the channel [1]. Therefore, passivation of the interface between gate oxide and Ge/III-V materials will require innovations to reach high device performances and EOT scaling. For this reason, a great technological effort is required to produce systems that yield the desired quality in terms of material purity, uniformity and interface control. Molecular Beam Epitaxy (MBE) has been shown to be an attractive technique to fabricate such devices, due to its potential to control at an atomic scale the *in-situ* deposition of the high-κ oxides and also the layer at the high-μ substrates interface [2,3]. In this work we investigate the passivation of n-Ge(001) and p-GaAs(001) using a common GeO$_2$ interfacial passivation layer (IPL) to reduce the defects at the interface.

EXPERIMENT

Experiments were carried out using a Riber 200 mm molecular beam epitaxy cluster production system. This unique multi chamber MBE system is composed of a MBE49 III-V growth reactor; a MBD49 high-κ oxide reactor, a passivation unit and a 25 holder's capacitance loading chamber. The high flexibility of current experimental setups enables us to perform various interfacial engineering schemes.

GeO$_2$ PASSIVATION OF Ge(001)

In a first part, the passivation of germanium will be discussed. Al$_2$O$_3$ has been chosen as gate oxide due to its κ-value of 10 and its large band offset discontinuities (> 2 eV) [4]. It is well known that the direct deposition of Al$_2$O$_3$ on Ge(001) does not provide a good passivation of the substrate [5]. Indeed, Al$_2$O$_3$ straight on Ge leads to the formation of an undesirable interfacial layer which presents a huge quantity of interface state density (D_{it}~5x10^{12} cm^{-2}eV^{-1}). In this case, an IPL is necessary in order to avoid any unstable layer formation, leading to the presence of traps at the Ge/high-κ interface. One solution proposed in this work is to form an *in-situ* controlled GeO$_2$ interlayer between Ge and the high-κ layer, to improve Ge passivation [5,6,7].

Prior to MBD, n-type Ge(001) wafers were chemically treated with NH$_4$OH/H$_2$O$_2$/H$_2$O followed by 2% HF dip. The Ge native oxide was then thermally desorbed *in-situ* by heating the substrate up to 700°C for 30 min. The GeO$_2$ interfacial layer was formed by exposing the clean Ge(001)-2x1 surface to an atomic oxygen flux at a low temperature of T~150°C. The Al$_2$O$_3$ thin films were elaborated by co-deposition of aluminium under an atomic oxygen flux of P(O$_{ato}$)~3.10^{-6} Torr. During the deposition, the substrate temperature was around T$_D$=250°C.

In order to study the structural quality of the n-Ge(001)/GeO$_2$/Al$_2$O$_3$ gate stack, a High Resolution Cross-Sectional Transmission Electron Microscopy (HR-TEM) image taken along <110>$_{Ge}$ of a stack with a 8.5 nm Al$_2$O$_3$ layer is shown in Fig. 1(a). The TEM picture evidences the good homogeneity of the stack and the absence of structural defects (like pinholes, surface etching ...) on the Ge surface. The Al$_2$O$_3$ thin film is completely amorphous and does not present any crystallites in the layer and presents a uniform thickness of 8 ± 0.5 nm and a relatively flat surface. The Ge/GeO$_2$-Al$_2$O$_3$ interface is sharp at the atomic-scale and continuous. X-ray Photoelectrons Spectroscopy (XPS) analysis is required to determine the chemistry at the interface. The Ge *3d* XPS spectra of a n-Ge(001)/GeO$_2$/Al$_2$O$_3$ heterostructure, is presented in Fig. 1(b). The exposure of the Ge surface to an oxygen plasma leads to the formation of completely oxidized GeO$_2$ ($\Delta BE_{(Ge-GeO2)}$ ~ 3.5 eV). But after the deposition of the Al$_2$O$_3$ oxide on the Ge(001)/GeO$_2$ heterostructure, a shift of the Ge-O peak to the lower energy ($\Delta BE_{(Ge-GeOx)}$ ~ 2.5 eV) reveals an impact of the high-κ deposition on GeO$_2$ layer. This behaviour suggests the change of the GeO$_2$ interfacial layer into germanium sub-oxides, GeO$_X$, or to the formation of aluminium germanate (AlGeO$_X$) at the high-κ/GeO$_2$ interface.

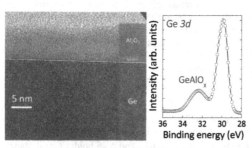 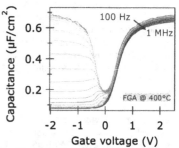

Fig.1: (a) TEM picture and (b) XPS Ge 3d spectra of a n-Ge(001)/GeO$_2$/Al$_2$O$_3$ heterostructure

Fig.2: C-V characterization of n-Ge(001)/GeO$_2$/Al$_2$O$_3$ gate stack

Figure 2 presents the C-V characteristic of n-Ge/1.2-nm-GeO$_2$/9-nm-Al$_2$O$_3$/Pt stacks, measured from inversion to accumulation and taken at different frequencies (100 Hz - 1 MHz). The capacitors were exposed to a FGA anneal at 400°C during 5 min. A well-behaved C-V curve is observed with very low frequency dispersion in accumulation and depletion as well as small flat-band voltage shift between the low and high frequency curves. No significant bump is observed, indicating that the D_{it} near mid-gap is small: D_{it}~2.10^{11}cm^{-2}.eV^{-1} and the surface Fermi level is likely unpinned [5,6].

Geo$_2$ PASSIVATION OF GaAs(001)

Gallium arsenide (GaAs), with its intrinsically superior electron mobility, has been considered as a good candidate for sub-15 nm node n-MOS [1]. The high D_{it} at the oxide/GaAs interface is the main origin of Fermi level pinning which disturbs the basic MOSFET-operation [8]. Several passivation techniques have been attempted to prevent Fermi level pinning: chalcogenide [9] or hydrogen [10] surface treatments, molecular beam epitaxy-grown Ga$_2$O$_3$(Gd$_2$O$_3$) oxide [11] or interfacial passivation layers such as a-Si [12] or a-Ge [13]. Although considerable improvements have been realized in D_{it} reduction, further developments are required to obtain high performance MOS devices.

In this contribution, we propose the use of a thin a-GeO$_2$ IPL to prevent GaAs-Fermi level pinning, using the same techniques developed for the passivation of Ge detailed elsewhere [5,6]. As the exposure of a crystalline Ge surface to an atomic oxygen plasma leads to the formation of a thin GeO$_2$ layer (with a self-limited thickness in function of the substrate temperature) and the ultra-low lattice mismatch between both GaAs and Ge semiconductors (f~-0.08%) allows defect free epitaxy of a Ge layer on a GaAs substrate, GeO$_x$ could be used as a passivation layer for p-GaAs(001).

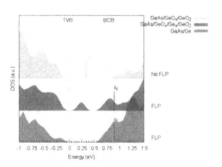

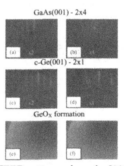

Fig. 3: DOS calculation of (a) GaAs(001)/c-Ge, (b) GaAs(001)/c-Ge/a-GeO$_X$ and (c) GaAs(001)/a-GeO$_X$ heterostructures

Fig. 4: RHEED patterns along the [110] and [1-10] azimuths of (a&b) the 2x4 p-GaAs(001) surface, (c&d) the 2x1 Ge epi-layer, (e&f) the a-GeO$_2$ layer after Ge oxidation

This possible road to passivate GaAs has then been investigated by Density Functional Theory [14] simulations. The density of states (DOS) calculations at the Local Density Approximation level (LDA) have been realized on three different systems: abrupt GaAs(001)/c-

Ge, GaAs(001)/c-Ge_2x1/a-GeO$_X$ and GaAs(001)/a-GeO$_X$ [Fig. 3]. These simulations show that the two systems containing crystalline Ge at the interface lead to the pinning of the Fermi level. But they also clearly reveal the possibility of using a thin amorphous layer made of GeO$_X$ to obtain an electrically unpinned gap. The major challenge resides in the control of the c-Ge thickness and the oxidation of this layer to avoid the diffusion of oxygen atoms at the Ge/GaAs(001) interface.

The GaAs layers were grown on p-type (001)-oriented GaAs epi-ready substrates. The GaAs substrates were heated to 580°C under As-rich conditions (P(As)=2.5×10^{-5} Torr), to remove the native oxide. The ~0.3μm-GaAs buffer layers [Be doped 5x10^{16} cm^{-3}] were grown at 580°C, at a growth rate of 1 ML.s^{-1} and under As-rich conditions (P(As)=1.8×10^{-5} Torr), leading to a (2×4) surface reconstruction during growth. Optimal epitaxial growth of pseudomorphic c-Ge thin film has been carried out under UHV (P~2×10^{-10} Torr) at a growth temperature of 425°C on the GaAs(001)-2x4 reconstructed surface (Fig. 4(a&b)). Such parameters are leading to a 2D growth of Ge which has been confirmed by the streak (2×1) reconstruction observed on the RHEED patterns (Fig. 4(c&d)). The 2x1 reconstruction is an indication of a monodomain Ge(001) growth replicating the initial stepped GaAs surface. The sample is then introduced into the high-κ chamber to perform a controlled oxidation of the Ge(001) surface. For this, the Ge/GaAs(001) heterostructure is exposed to an atomic oxygen flux (P(O$_{ato}$)=4×10^{-6} Torr) at 150°C. This step is leading to the formation of a completely amorphous GeO$_2$ layer on the surface of the GaAs(001) substrate as shown on the RHEED diagram of Fig. 4(e&f). The deposition of the high-κ dielectric, has been realized using the same conditions as previously described on Ge(001) substrate (§3).

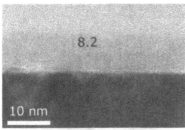

Fig.5: HR-TEM picture of a p-GaAs(001)/GeO$_2$/Al$_2$O$_3$ heterostructure

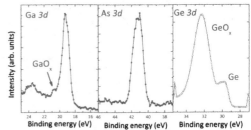

Fig.6: p-GaAs(001)/GeO$_2$/Al$_2$O$_3$ gate stack XPS core level spectrum of (a) Ga 3d, (b) As 3d and (c) Ge 3d

To study the structural quality of the p-GaAs/Al$_2$O$_3$ structure, TEM analyses have been performed. The cross-section HR-TEM image of the sample (shown in Fig. 5) shows that the interface isn't sharp at the atomic scale and presents some roughness at the GaAs/oxide interface. Energy Dispersive X-Ray (EDX) analyses at the GaAs substrate surface confirms that the bright and rough layer is Ge. These roughnesses could appear during the growth of crystalline Ge at the first stage of the epitaxy. But this is in contradiction with the streaks diffraction lines that are observed on the RHEED patterns. On this picture, the Al$_2$O$_3$ thin film shows a good homogeneity and thickness uniformity (8nm ± 0.5 nm), furthermore the high-κ oxide layer is completely amorphous and does not present any crystallites in the layer.

Investigation of the chemical properties of the p-GaAs/GeO$_2$/2nm-thick-Al$_2$O$_3$ heterostructure was performed by XPS. Fig. 6 summarizes the XPS core level spectra recorded at an incidence detection angle of 21.875°. The Ga *3d* core level spectrum, centred at 19.3 eV on Fig. 6(a), exhibits a very small feature on the higher energy level (~ 20.7 eV) which attests the presence of a very thin 0.4-nm-tick Ga-O layer at the interface. On figure 6(b), the peak centred at 41.7 eV is associated to the As *3d* core level. In this region, no presence of arsenic sub-oxides has been found. Finally, the Ge *3d* core level spectrum is shown on Fig. 6(c) and exhibits two peaks in this region. The weak peak centred at 29.8 eV corresponds to the Ge-Ge bonds from the crystalline Ge thin film. The strong peak at 32.3 eV is associated to the formation of an aluminium germanate (GeAlO$_x$) at the interface ($\Delta BE_{(Ge-GeOx)}$ ~ 2.5 eV). This analysis proves that the Ge layer isn't completely oxidized and a small amount of pure Ge is still present after the oxidation process step.

In order to study electrically the impact of the thin GeO$_2$ IPL, MOS capacitors have been realized on three different stacks: (A) p-GaAs(001)-(2x4)/9nm-Al$_2$O$_3$, (B) p-GaAs(001)/1.2nm-Ge-(2x1)/9nm-Al$_2$O$_3$ and (C) p-GaAs(001)/1.2nm-GeO$_2$/9nm-Al$_2$O$_3$. The thicknesses of the different stacks have been measured using XPS analysis. For that a top electrode, consisting in a 50-nm-thick Pt layer, was deposited *ex-situ* on each stack through a shadow mask and the back side Ohmic contact was formed using a composite stack of 30 nm of Au followed by 70 nm of AuZn. The Capacitance-Voltage (C-V) characterizations measured at room temperature, with the frequency in the range from 100Hz to 1MHz, of each as-deposited heterostructures are presented in Fig. 7. C-V characterizations of sample A, Al$_2$O$_3$ straight on GaAs(001), shows large frequency dependant flat-band voltage shift in accumulation and depletion. This is associated to the existence of interface states, probably due to the formation of interfacial defects during the exposure of the III-V surface to the atomic oxygen plasma. Sample B, with the thin crystalline Ge interlayer, clearly reveals poor electrical properties: the C-V characteristics exhibit huge frequency dispersion in accumulation and depletion. Such conduct clearly evidences the presence of states in high quantity at the interface, leading to the pinning of the Fermi level. This behaviour is in accordance with the previous *ab-initio* calculations. But in the case of sample C, with GeO$_x$ at the interface, the C-V curve displays a reduced frequency dispersion in accumulation and depletion compared to samples A & B. This indicates that the GeO$_2$ IPL demonstrates an important part in the passivation of the p-GaAs(001) surface.

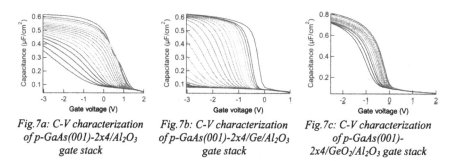

Fig.7a: C-V characterization of p-GaAs(001)-2x4/Al$_2$O$_3$ gate stack   *Fig.7b: C-V characterization of p-GaAs(001)-2x4/Ge/Al$_2$O$_3$ gate stack*   *Fig.7c: C-V characterization of p-GaAs(001)- 2x4/GeO$_2$/Al$_2$O$_3$ gate stack*

For GaAs-based structures, the interface is characterized by C-V measurements at 25°C and 150°C [8]. In the case of sample C, although the room temperature measurement gives promising electrical performances with low frequency dispersion, the high temperature measurement (not shown here) presents a very large density of interface states around the mid-gap energy. This points out that the Fermi level is still pinned in mid-gap. One hypothesis of this Fermi level pinning in the case of the p-GaAs(001)/a-GeO$_2$ system could be attributed to the existence of metallic Ge (non-oxidized Ge) which is still present between the dimer trenches of the (2x4)-GaAs(001) reconstructed surface. That could explain the origin of the Ge-Ge bonds observed at 28.9 eV on the Ge $3d$ XPS core level spectrum, indicating the presence of metallic Ge at the interface after the oxidation step.

CONCLUSIONS

The structural and electrical properties of a common GeO$_2$ IPL for n-Ge(001) and p-GaAs is promising for the application in high performance devices. On n-Ge(001), the GeO$_2$ IPL could provide an efficient passivation of the Ge interface with a very low D$_{it}$ value in mid gap and an unpinned surface. On p-GaAs(001), controlling of the different process steps (mono-crystalline growth of Ge and oxidation for the formation of the GeO$_2$ IPL) is a critical issue. In this work we have obtained straight electrical properties but the Fermi Level is still pinned in mid gap. The hypothesis of pinning origin associated to the presence of metallic Ge which is still present between the dimer trenches of the GaAs initial surface is actually also under investigation. The control of the different steps must be further addressed in order to optimize the passivation of GaAs.

ACKNOWLEDGMENTS

This work is part of the IMEC Industrial Affiliation Program on Ge/III-V devices and is supported by the European Commission's project FP7-ICT-DUALLOGIC no. 214579, "Dual-Channel CMOS for (sub)-22nm High Performance Logic". The authors would like to thank Dr. J. Villette from Riber, Dr. Th. Conard for XPS characterizations, J. Steenbergen for performing the anneals and H. Costermans for the metal gate deposition.

REFERENCES

1. A. Dimoulas et al., Advanced Gate Stacks for High Mobility Semiconductors, Springer (2007)
2. J.P. Locquet et al., J. Appl. Phys. **100**, 051610 (2006)
3. C. Merckling et al., Microelec. Eng. **84**(9-10), 2243 (2007)
4. C. Merckling et al., Appl. Phys. Lett. **89** (23), 232907 (2006)
5. F. Bellenger et al., ECS Trans. **16**(5), 411 (2008)
6. C. Merckling et al., Microelec. Eng., doi:10.1016/j.mee.2009.03.048 (2009)
7. H. Matsubara, T. Sasada, M. Takenaka, and S. Takagi, Appl. Phys. Lett. **93**, 032104 (2008)
8. G. Brammertz et al., Appl. Phys. Lett. **93**, 183504 (2008);
9. J. Massies, F. Delazy, N.T. Linh, J. Vac. Sci. Technol. **17**(5) 1134 (1980)
10. A. Callegari, P.D. Hoh, D.A. Buchanan, D. Lacey, Appl. Phys. Lett. **54**(4), 332 (1989)
11. R. Droopad et al., J. Vac. Sci. Technol. B **24**, 1479 (2006)
12. P. de Souza et al., Appl. Phys. Lett. **92**, 153508 (2008);
13. H.-S. Kim et al., Appl. Phys. Lett. **92**, 032907 (2008)
14. J.M. Soler et al., J. Phys. Condens. Matter. 14, 2745 (2002)

Mater. Res. Soc. Symp. Proc. Vol. 1155 © 2009 Materials Research Society   1155-C02-04

Material Properties, Thermal Stabilities and Electrical Characteristics of Ge MOS Devices, Depending on Oxidation States of Ge Oxide: Monoxide [GeO(II)] and Dioxide [GeO₂(IV)]

Yoshiki Kamata[1], Akira Takashima[2], and Tezuka Tsutomu[1]
[1]MIRAI-Toshiba, 1, Komukai-Toshiba-cho, Saiwaiku, Kawasaki, 212-8582, Japan
[2]Toshiba Corporation, 1, Komukai-Toshiba-cho, Saiwaiku, Kawasaki, 212-8582, Japan

ABSTRACT

Ge monoxide [GeO(II)] and dioxide [GeO₂(IV)], which are selectively formed on Ge substrate by controlling pH and redox potential in pretreatment solution, have been confirmed by XPS. ΔE_C in GeO(II)/Ge and GeO₂(IV)/Ge are almost the same, whereas ΔE_V in GeO(II)/Ge is smaller than that in GeO₂(IV)/Ge, resulting in smaller E_g of GeO(II). GeO(g) desorption is suppressed in LaAlO₃/Ge gate stack, whereas GeO(g) desorbs through LaAlO₃ layer when there is an intentional interfacial GeO(II) layer, leading to a large increase in J_g. GeO(g) desorption temperature in Ge oxide/Ge gate stacks decreases with the increase in the ratio of GeO(II) in Ge oxide and is independent of the oxidation techniques. Since GeO(g) desorption is accompanied by H₂O(g) desorption, a new model to explain the GeO(g) desorption phenomena is proposed, in which Ge(OH)₂ decomposes into GeO(g) and H₂O(g). Highly effective etching methods of Ge oxide, using HCl solution and HCl vapor at higher temperature than boiling point of Ge (hydro-)chloride have been demonstrated.

INTRODUCTION

High-k/Ge MISFETs are very promising for future nanoscale LSIs[1, 2]. The most important technical issue concerning high-k/Ge gate stacks is the passivation of the Ge surface, or the control of the interfacial Ge oxide. Although Ge oxide consists of GeO(II) and GeO₂(IV) [1, 2], little has been reported about influences of GeO(II), distinguishing them from those of GeO₂(IV), on thermal stability and electrical characteristics of Ge MOS devices.

In this study, we investigate material properties, thermal stabilities and electrical characteristics of Ge MOS devices, depending on the type of Ge oxide: GeO(II) and GeO₂(IV). In particular, we report on (i) a selective formation of GeO(II) or GeO₂(IV) that is confirmed by XPS, (ii) band alignments of GeO(II)/Ge and GeO₂(IV)/Ge, (iii) relationship between J_g and GeO(g) desorption in LaAlO₃/Ge gate stacks with or without intentional interfacial Ge oxide, (iv) universality of GeO(g) desorption depending on GeO(II) ratio in Ge oxide and independent of oxidation technique, (v) a new GeO(g) desorption model explaining GeO(g) desorption accompanied by H₂O(g) desorption, and (vi) highly effective etching methods of Ge oxide using HCl solution and HCl vapor at higher temperature than boiling point of Ge (hydro-)chloride.

EXPERIMENTAL

LaAlO₃ film of 10nm was deposited at 200°C by using MBE apparatus on Ge substrates (Sb, ~0.3Ωcm) with or without intentional interfacial Ge oxide layer. X-ray photoelectron spectroscopy (XPS) analyses were performed with monochromatic-Al Kα source. Thermal desorption spectroscopy (TDS) was performed for LaAlO₃/Ge gate stacks and Ge oxide/Ge gate stacks. MOS capacitors were subjected to a post-deposition annealing (PDA) in N_2 at 400°C for 30 min prior to gate electrode formation.

RESULTS & DISCUSSION

The main topics of this study are investigations of an influence of GeO(g) desorption on gate leakage current density (J_g), physical reasons for GeO(g) desorption phenomena, and solutions for suppressing GeO(g) desorption. To discuss these topics, material properties of Ge oxide are also investigated in detail, distinguishing GeO(II) and GeO$_2$(IV).

(i) Selective formation of GeO(II) and GeO$_2$(IV) on Ge

A Pourbaix diagram of Ge indicates that GeO(II) and GeO$_2$(IV) can be selectively formed on Ge substrate by controlling pH and redox potential in pretreatment solution[3]. With XPS measurement, we confirmed that when Ge substrate was pretreated, for example, by HF(20%)+H$_2$O$_2$(0.1%) solution, GeO(II) was formed on Ge (figure 1). It is noted that color of Ge surface changed from colorless to brown[4], which is not due to fringe but to the material property of GeO(II), since the brown color was independent of watching angle (figure 2). Moreover, GeO(II) can be oxidized to colorless GeO$_2$(IV) by H$_2$O$_2$, which was also confirmed by XPS measurement (figure 1).

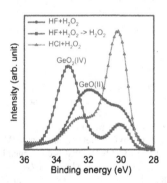

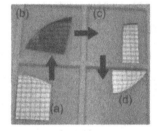

Figure 2 Color of Ge surface of (a) as-received, (b) GeO(II), (c) GeO$_2$(IV), and (d) after water rinse for GeO$_2$(IV).

Figure 1 XPS Ge 3d spectra for GeO(II) and GeO$_2$(IV), corrected using C1s of 284.8eV. TOA=15deg. The spectra of Ge substrate pretreated by HCl with H$_2$O$_2$ is also shown as reference

These selective formation techniques of Ge oxide, in particular GeO(II), enable us to investigate the influences of each Ge oxide on thermal stability and electrical characteristics of Ge MOS devices, distinguishing GeO(II) from GeO$_2$(IV), as explained in the following.

(ii) Band alignment in GeO(II)/Ge and GeO$_2$(IV)/Ge

For example, we can investigate the difference in band alignments between GeO(II)/Ge and GeO$_2$(IV)/Ge. ΔE_C in GeO(II)/Ge and GeO$_2$(IV)/Ge are almost the same, whereas ΔE_V in GeO(II)/Ge is smaller than that in GeO$_2$(IV)/Ge, resulting in smaller E_g of GeO(II) (figures 3~5).

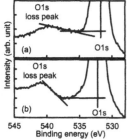
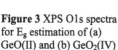
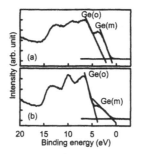
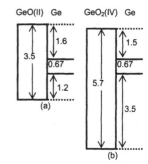

Figure 3 XPS O1s spectra for E_g estimation of (a) GeO(II) and (b) GeO$_2$(IV)

Figure 4 XPS valence spectra for ΔE_V estimation of (a) GeO(II) and (b) GeO$_2$(IV)

Figure 5 Band alignment for (a) GeO(II)/Ge, and (b) GeO$_2$(IV)/Ge

Although the difference of ΔE_V can be expected to affect J_g, several orders of magnitude increase in J_g in the case of intentional formation of interfacial Ge oxide in high-k/Ge gate stack independent of gate voltage polarity is not due to ΔE_V but to GeO(g) desorption as explained in the next subsection.

(iii) Influence of GeO(g) desorption through LaAlO$_3$ layer on J_g

It has already been reported that GeO(g) desorption through high-k film degrades J_g several orders of magnitude in HfO$_2$/Ge gate stack[1, 2]. In this study, we investigated the correlation between GeO(g) desorption from LaAlO$_3$/Ge gate stack and J_g, and the influence of interfacial Ge oxide layer on the relationship.

GeO(g) desorption was suppressed in LaAlO$_3$/Ge gate stack, whereas GeO(g) desorbed through LaAlO$_3$ layer when there was an intentional interfacial GeO(II) or GeO$_2$(IV) layer, leading to a large increase in J_g. (figure 6). The degradation of J_g in the case of GeO(II) interlayer was more severe than that of GeO$_2$(IV). Since GeO(g) in LaAlO$_3$/GeO(II)/Ge gate stack started to desorb at below PDA temperature of 400°C, we considered GeO(g) was the origin of the increase in J_g. Although the desorption temperature (T_d) of GeO(g) in GeO$_2$(IV) sample was higher than that in GeO(II) one, J_g in GeO$_2$(IV) was large enough compared to that of the sample without interlayer. As explained in the following, this GeO$_2$(IV) sample contained a little GeO(II) that is expected to lead to the increase in J_g. The increase in J_g originates in the reducing nature of GeO(g) which reduces high-k layer when GeO(g) desorbs through high-k film[1].

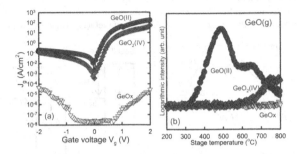

Figure 6 (a) J_g of LaAlO$_3$/GeO(II)/Ge, LaAlO$_3$/GeO$_2$(IV)/Ge, and LaAlO$_3$ /GeOx /Ge. GeO(II) is formde by HF with H$_2$O$_2$ pretreatment. GeO$_2$(IV) is formed by plasma oxidation. GeOx is formed by HCl pretreatment at 110°C followed by H$_2$O rinse. (b) TDS spectra of ^{74}Ge^{16}O (M/Z =90) from the samples in (a).

With intentional interfacial GeO(II) layer, GeO(g) desoption degrades J_g several orders of magnitude. GeO(g) desorption depends not only on the type of high-k material[1] but also on the type of interfacial Ge oxide.

(iv) GeO(g) desorption from Ge oxide/Ge gate stacks

To investigate GeO(g) desorption phenomena in detail, we studied GeO(g) desorption from Ge oxide/Ge gate stacks prior to high-k film deposition (figures 7, 9). Many types of Ge oxide were prepared by different oxidation techniques: chemical oxidation with pretreatment solution, thermal oxidation and plasma oxidation, leading to different binding energy of Ge oxide, which means different ratio of GeO(II) in Ge oxide (figures 1, 8, 10).

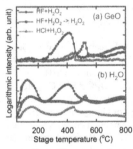

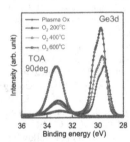

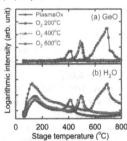

Figure 7 TDS spectra of (a) ^{74}Ge^{16}O (M/Z =90) and (b) H$_2$O for the samples in figure 1.

Figure 8 XPS Ge3d spectra from samples oxidized in O$_2$ ambient at 200, 400, 600°C and by O$_2$ plasma for 30 min, corrected using C1s of 284.8eV.

Figure 9 TDS spectra of (a) ^{74}Ge^{16}O (M/Z =90) and (b) H$_2$O from samples in figure 10.

172

Figure 10 indicates the universal relationship between T_d and GeO(II) ratio such that T_d depended on only GeO(II) ratio and was independent of oxidation methods. Moreover, T_d decreased with the increase in GeO(II) ratio. It is noted that T_d from GeO(II)/Ge gate stack was lower than not only sublimation temperature (700°C) but also previously reported temperature (~ 400°C) [5]. The most important finding is that all GeO(g) desorption was accompanied by $H_2O(g)$ desorption at the same temperature and that an amount of GeO(g) desorption was proportional to that of $H_2O(g)$ desorption (figures 7, 9). Previously, GeO(g) desorption was thought to occur by the reaction: $GeO_2(IV) + Ge(s) \rightarrow 2GeO(g)$[5]. To explain our experimental results, we add a new model (figure 11) in which Ge hydroxide [Ge(OH)$_2$] decomposes into GeO(g) and $H_2O(g)$. If GeO(II) does not exist in Ge oxide, which means pure $GeO_2(IV)$, GeO(g) desorption is suppressed unless GeO(II) is produced by the reaction: $GeO_2(IV) + Ge(s) \rightarrow 2GeO(II)$[5]. When both GeO(II) and H_2O exist, once Ge(OH)$_2$ is formed, Ge(OH)$_2$ cracks into GeO(g) and $H_2O(g)$. Since GeO(g) desorbs as a result of cracking of Ge(OH)$_2$, T_d is independent of sublimation point.

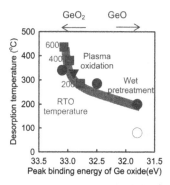

Figure 10 T_d universal curve (gray line) for Ge oxide on Ge substrate. It is noted that many T_d (*e.g.* open circle) are defined in the wet pretreated Ge oxide.

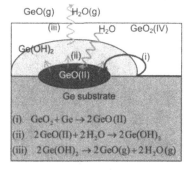

Figure 11 Schematic figure shows a model for GeO(g) desorption accompanied by $H_2O(g)$ desorption.

Some reports have recently pointed out that an interface between Ge substrate and GeO_2 layer oxidized by strong oxidation condition (~550°C or high pressure) exhibits superior interface properties [6, 7]. In light of our experimental results, their good electrical characteristics are thought to come from suppression of GeO(II) formation and GeO(g) desorption. Since $GeO_2(IV)$ in figures 6 and 10 was formed by plasma oxidation, it contained a little GeO(II). This $GeO_2(IV)$ was comparable to thermal oxidation at 400°C that was lower than the above strong oxidation temperature of 550°C. Therefore, GeO(g) desorbed at the interface and J_g degraded. Suppression of GeO(II) formation is a key issue for suppressing an increase in J_g and sustaining gate stack integrity.

(v) Chemical etching of Ge oxide using HCl solution at high temperature and HCl vapor

Although strong oxidation condition is attractive for obtaining good interface, complete absence of Ge oxide is likely to be necessary for achieving the thin EOT below 0.5nm required in ITRS, since permittivity of GeO_2 is as low as 7. Furthermore, Ge oxide on Ge decomposes or desorbs, leading to degradation of the gate stack integrity. Therefore, we investigated chemical etching processes of Ge oxide in the pretreatment. Ge oxide remains on Ge substrate after conventional pretreatments[8]. If Ge oxide is etched at once, Ge surface is reoxidized at deionized water (DIW) rinsing process[9]. It is important to determine whether residual Ge oxide is old initial oxide layer or fresh reoxidized one, since initial Ge oxide may contain metal contamination such as Cu or Fe, degrading electrical characteristics[8]. Initial oxide layer must be removed once, but GeO(II) has a resistance against high concentration HF solution or HCl one (figure 12), which is reported to be effective for etching Ge oxide[9, 10]. Since each pretreatment in figure 12 is followed by DIW rinse, GeO(II) is also insoluble. Since the brown color does not disappear, the oxide is not due to reoxidation but to residual one. Although one of the etching methods is using oxidizing agents, H_2O_2 (figure 12), H_2O_2 also oxidizes Ge, leading to an increase in surface roughness. Therefore, in this study we investigated another two etching processes of GeO(II) as well as GeO_2(IV).

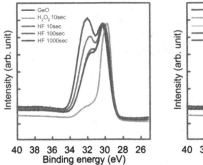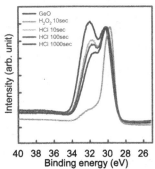

Figure 12 XPS Ge3d spectra for GeO(II)/Ge pretreated by HCl(20%), HF(20%) or H_2O_2(0.1%) followed by DIW rinse. Energy shift is corrected using C1s of 284.8eV.

One is a pretreatment in HCl solution [10, 11] at higher temperature than boiling points of Ge (hydro-)chlorides: $GeCl_4$ of 83°C and $GeHCl_3$ of 75°C. In HCl solution, Ge wafer starts hovering at above ~80°C with bubble that is considered to be (hydro-)chlorides (table 1), leading to a sharp decrease in Ge oxide (figure 13). With this HCl pretreatment, GeO(g) desorption and the increase in J_g are suppressed (figure 6). The other is HCl vapor pretreatment. Firstly, we analytically calculated thermal reactivity of Ge oxide, and SiO_2 as a reference, with gas phase hydrogen halides (HX(g), X=F, Cl, Br and I) using Gibbs free energy (figure 14). SiO_2 can be etched only by HF vapor, whereas both Ge oxide (GeO and GeO_2) can be etched by all hydrogen halide gases. In practice, HF(g)[12] and HCl(g) are attractive. Figure 15 shows that Ge oxide can be actually etched by HCl vapor. HCl vapor at higher temperature also leaded to smaller amount of Ge oxide. It is noted that a series of HCl pretreatment terminated Ge surface with Cl[11] that is reoxidized at DIW rinse process (figure 13).

Table 1 Boiling point of Ge (hydro-) chloride (GeCl₄ or GeHCl₃), and its producing reaction from Ge oxide with HCl.

Gas	Boiling point ($^{\circ}$C)	$GeO_2(IV)$	$GeO(II)$
$GeCl_4$	83	$GeO_2 + 4HCl \rightarrow GeCl_4 + 2H_2O$	$GeO + 4HCl \rightarrow GeCl_4 + H_2O + H_2$
$GeHCl_3$	75	$2GeO_2 + 6HCl \rightarrow 2GeHCl_3 + 2H_2O + O_2$	$GeO + 3HCl \rightarrow GeHCl_3 + H_2O$

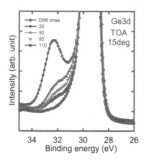

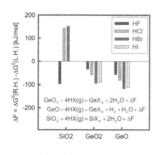

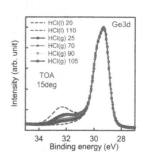

Figure 13 XPS Ge3d spectra for Ge substrate after HCl pretreatment at 20, 50, 95, 110°C, and 110°C followed by water rinse, corrected using Ge-Ge of 29.3eV.

Figure 14 Difference of Gibbs free energy of oxide etching reaction with halogen acid gas.

Figure 15 XPS Ge3d spectra for Ge substrate after HCl(g) pre-treatment at 25, 70, 90, and 105°C, corrected using Ge-Ge of 29.3eV. XPS Ge 3d spectra pretreated by HCl(l) at 20 and 110°C (figure 13) are also shown as references.

SUMMARY & CONCLUSION

Material properties, thermal stabilities and electrical characteristics of Ge MOS devices are revealed, distinguishing differences between two kinds of Ge oxide: GeO(II) and GeO₂(IV). Selective formations of GeO(II) and GeO₂(IV) with pretreatment solution controlling pH and redox potential are confirmed by XPS. ΔE_C in GeO(II)/Ge and GeO₂(IV)/Ge are almost the same, whereas ΔE_V in GeO(II)/Ge is smaller than that in GeO₂(IV)/Ge, resulting in smaller E_g of GeO(II). GeO(g) desorption is suppressed in LaAlO₃/Ge gate stack, whereas GeO(g) desorbs through LaAlO₃ layer when there is an intentional interfacial GeO(II) layer, leading to a large increase in J_g. The GeO(g) desorption temperature decreases with the increase in the ratio of GeO(II) in Ge oxide and is independent of the oxidation techniques. Since GeO(g) desorption is accompanied by H₂O(g) desorption, a new model to explain the GeO(g) desorption phenomena is proposed, in which Ge(OH)₂ decomposes into GeO(g) and H₂O(g). Highly effective etching methods of Ge oxide, using HCl solution and HCl vapor at above boiling point of Ge (hydro-)

chloride, are demonstrated. Controlling Ge oxide, in particular GeO(II), is a key issue for the fabrication process of high-k/Ge MOS devices.

ACKNOWLEDGMENT
This work was supported by NEDO.

REFERENCES
1.   Y. Kamata, Materials Today **11**, 30-38 (2008).
2.   Y. Kamata, Tutorial H, Materials Research Society Spring Meeting 2008
3.   M. Pourbaix, "*Atlas of Electrochemical Equilibria in Aqueous Solutions*". (Pergamon Press, London, 1966).
4.   N. N. Greenwood and A. Earnshaw, "*Chemistry of the Elements*", 2nd ed. (Pergamon, Oxford, 1997).
5.   K. Prabhakaran, F. Maeda, Y. Watanabe and T. Ogino, Appl. Phys. Lett. **76**, 2244 (2000).
6.   H. Matsubara, T. Sasada, M. Takenaka and S. Takagi, Appl. Phys. Lett. **93**, 32104 (2008).
7.   T. Nishimura, C. H. Lee, K. Kita, K. Nagashio and A. Toriumi, Extended Abstract of 2008 International Workshop on Dielectric Thin Film for Future ULSI Devices (IWDTF), 81 (2008).
8.   Y. Kamata, T. Ino, M. Koyama and A. Nishiyama, Appl. Phys. Lett. **92**, 63512 (2008).
9.   M. Sakuraba, T. Matsuura and J. Murota, Mater. Res. Soc. Symp. Proc. **535**, 281 (1999).
10.   B. Onsia, T. Conard, S. D. Gendt, M. Heyns, I. Hoflijk, P. Mertens, M. Meuris, G. Raskin, S. Sioncke, I. Teerlinck, A. Theuwis, J. V. Steenbergen and C. Vinckier, Diffus. Defect Data B, Solid State Phenom. **103-104**, 27 (2005).
11.   S. Sun, Y. Sun, Z. Liu, D.-I. Lee, S. Peterson and P. Pianetta, Appl. Phys. Lett. **88**, 21903 (2006).
12.   C. O. Chui, S. Ramanathan, B. B. Triplett, P. C. McIntyre and K. C. Saraswat, IEEE Electron Device Lett. **23**, 473 (2002).

177

Printed in the United States
By Bookmasters